W9-AZB-323

One-Minute Readings

Issues in Science,
Technology, and Society

Richard F. Brinckerhoff
Department of Science
Phillips Exeter Academy
Exeter, New Hampshire

▲▼ **Addison-Wesley Publishing Company**
Menlo Park, California • Reading, Massachusetts • New York
Don Mills, Ontario • Wokingham, England • Amsterdam • Bonn
Sydney • Singapore • Tokyo • Madrid • San Juan • Paris
Seoul, Korea • Milan • Mexico City • Taipei, Taiwan

This book is published by the Addison-Wesley Innovative Division.

Managing Editor: Michael Kane
Project Editor: Mali Apple
Editorial Development: Linda Poderski
Design: Jeff Kelly
Cover Art: Rachel Gage

ISBN 201-23157-3

7 8 9 10 11 12 - EB - 00 99 98 97 96

Contents

Titles marked with an asterisk (*) duplicate titles earlier in the table.

Introduction

This book is different from most of your other school books. Yes, it is full of questions, but:

— none of the questions have any correct answers. Your teacher doesn't have any either.

— all of the questions ask your opinion and are open to argument among your classmates, friends, or family.

— most of the questions will give you a new "slant" on the science course you are taking.

— most of the questions deal with "hot" issues you are learning about through the newspaper, magazines, and TV.

In writing this book, I looked for places where the science you are studying now has something important to do with the world, with the USA, and with your hometown daily life. In most of those places, applications of science are raising tough questions and are creating problems that science cannot answer.

Instead, the answers are being worked out right now by the actions of voters, taxpayers, parents, and consumers. Whether you like it or not, these actors include *you*. This book is intended to give you some practice—a warm-up if you will—for the kind of decisions you will have to make often as you experience life.

To deal with the questions in this book, you will need a knowledge of science to find not the "right" answers (because there are none) but the "best possible" answers. Sometimes finding a "best possible" answer will be frustrating. You may need to argue with others whose "best possible" answer differs from yours. You may never know how your answer would work if it were tried. It will often be fun. It will always be useful practice, for that is precisely the kind of problem solving that parents, politicians, business people, and, yes, scientists, have to do all the time.

Welcome to the real world!

Biology

We hold these truths to be sacred and undeniable; that all men are created equal and independent, that from that equal creation they derive rights inherent and inalienable, among which are the preservation of life, and liberty, and the pursuit of happiness.

— THOMAS JEFFERSON

Issue 1

Definition of "Life"

Biologically, when does a cat's life begin? At conception? At birth? Somewhere between? Biologically, at what moment did your life begin? At the same stage as a cat's? Legally, at what moment did your life begin?

Does a legal definition establish a parent's right to the abortion of (a) a healthy but unwanted child? (b) a child proved to have a serious incurable defect? When do parents' moral obligations begin—at the beginning of life as defined by biology or by the law?

Consider the recent case of an anencephalic newborn baby (born missing all or part of the brain). The doctors (with the parents' consent) removed the heart for another child in critical condition, awaiting a heart transplant. Even though the baby with anencephaly still had brain wave patterns (and was, therefore, still alive?), no anencephalic child has ever lived beyond a few days. The question is: Was this murder or not?

Biologically, at what moment does death occur? When breathing stops? When the heart stops beating? When brain waves cease? Somewhere between? Legally, at what moment is someone considered to be dead? Are the legal and biological moments the same? Does this legal definition establish a family's (or a doctor's) right to end the life of a victim dying slowly of a painful, incurable disease?

Think about this . . . *If the aging process could be significantly slowed to allow people to live two or three centuries, would you want to live that long?*

Issue 2

Dollar Value of a Human Life

In dollars, how much money is your life worth? Or your neighbor's? Is your answer determined by a human being's responsibilities? By his or her contribution to the economy or to society? By sentiment? This is not just a theoretical question.

In the process of transforming coal into coke for use in making iron and steel, a mass of coal is heated to over 1100°C (2000°F) for up to 16 hours. The coal gives off hydrocarbon gases known to cause lung cancer. The remedy is clear and effective—engineering improvements and changes in work practices that steel industry experts say would cost about $200 million every year. The changes would save an estimated 109 lives a year plus untold human suffering.

At a hearing of the Occupational Health and Safety Administration (OSHA) considering such remedies, it was argued that the remedies would cost between $1 million and $9 million per worker saved. Since each worker was destined to earn only $200,000 to $500,000 in his or her career, the cost of the controls outweighed the benefits and, it was argued, nothing should be done.

For another example, if it would cost auto manufacturers $90 million to install improved bumpers that will save 100 lives per year, is it worth doing? If each life is valued at $1 million, that is a saving of $10 million. Therefore, on a strict cost-accounting basis, it is worth it.

In theory such an approach helps industries, government agencies, and the courts decide issues affecting loss of human life. For example, the Environmental Protection Agency (EPA) puts the value of a human life at between $475,000 and $8.3 million. The Federal Aviation Agency (FAA) assumes $1 million, while the Consumer Product Safety Commission puts it at $2 million.

If you are horrified by these facts, how do you decide what a human life is worth when health and safety legislation is being written or legal damages are being determined? As a regulator or a judge, you *would* have to decide, *even if your decision would be to do nothing.* As examples of cases requiring decisions, consider whether it would be cost effective:

— to maintain crash fire trucks at small airports.

— to prolong the expensive safety testing of a new drug.

— to pay for the regular inspection of the nation's 565,000 highway bridges, half of which (according to the National Highway Traffic Safety Administration) are potentially unsafe.

— to close a copper smelting factory (in Tacoma, Washington, in 1983) with the certain loss of 800 jobs, or to accept the uncertain but demonstrated risk of lung cancer from arsenic in the air, resulting in about 1 death per year.

— for an airline to lower the age of retirement of the airliners in its fleet when one crash occurs.

Issue 3

Debating Dissection

Should a biology student be *required* to take part in animal dissections?

In a case in New Jersey, a 17-year-old high school student refused to dissect a cat in biology class. In defense of his stand, he said, "I think dissection just reinforces the idea that animal life is cheap. I feel it's an inherently objectionable thing to do."

But a past president of the New Jersey Science Teachers Association held an opposite view. He argued that dissection has importance and value and that no textbook could compare with the hands-on experience dissection offers.

How would you have resolved this conflict?

Think about this . . . If you could have known in 1942 what the release of atomic energy was going to mean to the world since then, pro and con, would you have voted to unleash the first nuclear fission reaction in Chicago?

You may disapprove of South Africa's abuse of human rights, but South Africa is the source of 81 percent of the world's supply of manganese, which is necessary for the hardening of steel. Should we buy from South Africa?

Issue 4

Daily Hazards to Life and Health

When you *know* some things can harm you, how do you choose among risks to your life and health?

The general public smokes billions of cigarettes a year while banning an artificial sweetener because of a one-in-a-million chance that it might cause cancer. The same public eats meals full of fat, flocks to cities prone to earthquakes, and goes hang gliding while it frets about pesticides in foods, avoids the ocean for fear of sharks, and breaks into a cold sweat on airline flights.

Do any of these actions make good sense?

By broad statistical measures, Americans never have been safer. Life expectancy at birth in 1986 was 74.8 years, up a full 4 years since 1970, largely because of a dramatic drop in deaths from heart disease and strokes.

Coal-generated electricity costs thousands of lives a year through mining and transportation accidents, and illness due to pollution, yet a majority of the public prefers it to nuclear-generated electricity. Should the American public have a say about the risk of nuclear power, which has claimed, by one estimate, only three lives in accidents in the last 30 years and yet does not have the sense to wear seat belts in automobiles, which regularly kill close to 45,000 Americans every year?

Do you base decisions about such matters solely on the number of deaths, injuries, and damage, or are there other important but hard-to-quantify feelings and values you need to consider in making a risk decision? If so, when should they prevail? (See "Risk," p. 120.)

Trying to measure the risks from various presumed hazards to daily life, Richard Wilson, a physicist and acting director (1980) of the Energy and Environmental Policy Center at Harvard University, used the best evidence available at that time to calculate the level of exposure that increases your chance of death by 1 part in 1 million (which reduces your life expectancy by 8 minutes). His findings:

From cancer:
Smoking 1.4 cigarettes
Living 2 months with a cigarette smoker
One X ray in a good hospital
Eating 100 charcoal-broiled steaks
Eating 40 tablespoonfuls of peanut butter
Living 20 years near a polyvinyl chloride plant
Living 150 years within 20 miles of a nuclear power plant
Crossing the ocean by air (from cosmic rays)
Visiting for 2 months in Denver (from cosmic rays)

From accidents:
Traveling 1000 miles by air
Traveling 150 miles by automobile
Working 3 hours in a coal mine
Rock climbing for 1.5 minutes

(Richard Wilson, "Staying Alive in the 20th Century," *Science 85*, October 1985, p. 31–39.)

Think about this . . . *Let's explode a nuclear device on the moon or an asteroid to find out what's inside the celestial body. Good idea? Why?*

Issue 5

Tobacco

Nicotine, the drug in tobacco, is as addictive as cocaine or heroin. Tobacco claims 4 lives every minute in the U.S. (350,000 per year), which is more Americans than were killed in all of World War II. Illegal drugs, by contrast, kill an estimated 10,000 annually, although the social costs are immeasurably higher. Worldwide, smoking kills about 2.5 million people each year.

Ironically, while the U.S. contemplates imposing sanctions on Latin American countries that export cocaine and heroin to us, our government and the cigarette industry are collaborating to increase exports of our own drug, tobacco, to them!

Should tobacco companies be free to advertise their wares without restriction here and abroad? Remember that cigarette sales abroad are a significant factor in reducing our trade deficit. Would you approve of restrictive laws if it were shown that somehow tobacco was killing baby seals?

(See "Pesticide Pollution," p. 28.)

Think about this . . . Science is an important component of much of the legislation before the U.S. Congress each year. Should Congressmen and Congresswomen be required to pass some high school level science courses before they take office?

Issue 6

Drugs and Sports Technology

Is winning worth it?

As sports medicine and sports technology become more sophisticated, they improve an athlete's performance. Do they also devalue it? Certainly they raise questions. For example:

— Is there a moral difference between taking energy-giving glucose pills and taking an anabolic steroid that increases muscle mass in order to improve performance?

— If drug use is immoral, what about taking a drug that deadens the pain of an injured foot? If that's okay, what about taking a drug that does nothing but increase aggressiveness?

— If introducing foreign substances into the body is wrong, what about the Olympic cyclist who, in mid-race, had a reinfusion of his *own* blood that had been drawn several weeks earlier when he was "fresh" and his red cell count was much higher?

— If attempts to alter the body's chemistry are immoral, what about taking vitamins during training and invoking the science of nutrition to suggest special diets?

If it was wrong to try to alter the athlete's body chemistry, what about improving the equipment? For example:

— Graphite- and boron-based tennis racquets and fishing rods are far lighter and stronger than earlier styles.

— Lightweight body protection for football players has reduced the rate and severity of injuries.

— Nylon rope and new strong, lightweight alloys are making rock climbing safer and making possible more difficult climbs.

How does your school athletic coach respond to all the above questions?

Note that people climbing Mt. Everest commonly use oxygen. At least half of the 9000 athletes who competed at the 1988 Olympics in Seoul, Korea, used anabolic steroids in training, according to estimates by medical and legal experts.

Think about this . . . *Should people who carry antibodies to the AIDS virus but have not yet developed the disease be isolated or fired from their jobs?*

Issue 7

Patenting Plants and Animals

It is possible now to invent and patent new plants and animals—if you can afford it.

In April 1988, eight years after the U.S. Supreme Court legalized the patenting of genetically altered microorganisms, the U.S. Patent Office ruled that genetic engineers may now patent *higher* life forms—even mammals. One immediate result may be that, with patents at stake and many millions of dollars invested in competitive research and development, scientists developing new organisms for sale are unlikely to share information freely with colleagues, a subversion of one of the important characteristics of good science.

Another bad result, critics charge, is that competitive pressures to win profitable patents may encourage cruel animal experiments and may also wipe out small farmers unable to afford genetically improved livestock. (See the first following example.) They worry, too, that escaped or mutant organisms might unleash a biological catastrophe of some sort on an unprepared world.

On the other hand, supporters of genetic engineering insist that genetic manipulation may

— hasten a cure for AIDS and other diseases
— help reduce chemical pollution of the environment
— produce new nonpolluting substances to replace fossil fuels as a source of energy and industrial raw materials
— produce new agricultural food plants.

Support or disapproval of these conflicting positions must take into account unresolved ethical and safety questions. For example:

— Is it safe to release gene-altered plants or animals into the environment?
— Is it ethical to alter the genetic structure of animals? Do animals have rights?

— By what rules or standards does our society decide whether a new technology should be pursued?

Dorothy Nelkin, a professor in Cornell University's Program on Science, Technology, and Society, says the issues run from the environment to religious thought. Billions of dollars, hundreds of thousands of jobs, and America's ability to compete in the world's agricultural and pharmaceutical marketplace are at stake. Only obtaining a patent justifies the tremendous expense of the research.

As an example of what may happen, consider BGH, a synthetic protein identical to a hormone produced naturally in a cow's pituitary gland. Through genetic engineering, researchers have developed a bacterium that can produce the BGH protein in commercial quantities. Cows that were given the protein in daily injections increased their milk yields by 10 to 20 percent. The U.S. already is producing more milk than it knows what to do with. With no market for more milk, an increase in milk productivity will require a decrease in the number of cows in production—which will result in a corresponding decrease in the number of dairy farms. If the Food and Drug Administration (FDA) approves the use of BGH, many small farms may be driven out of business.

This is only one example of how genetic engineering might speed the elimination of small farmers and increase the size and industrialization of farms all over the world. Already the number of farms in the U.S. has dropped from 4.8 million (in 1954) to 2.3 million (in 1985). The average size of the surviving farms has more than doubled from 98 hectares (242 acres) to about 180 hectares (445 acres). If this decline continues, what effect will it have on *your* community? On local and state taxes? On food prices? On the social diversity of urban and rural society as ever more farmers go out of business? Should failing farmers receive financial assistance? Does the recent widespread failure of savings and loans in our country bear on these questions? Should the use of BGH be made illegal?

In April 1990, Wisconsin banned the sale or use of BGH in the state for at least 1 year in an effort to protect the state's dairy farmers and rural communities. Is this ban comparable to the ban that the city council of Cambridge, Massachusetts, placed on some kinds of research in recombinant genetics in January 1977?

As a second example, fisheries biologists at Auburn University in Alabama are isolating the gene in catfish that produces growth hormone. Their work shows promise of reducing the time needed for catfish to grow big enough to be eaten as food improving the vitality of the South's $300 million aquaculture industry. A plausible next step is to alter fish genetically to survive in polluted lakes and rivers.

Yet another example is a genetically altered strain of rice with greater drought tolerance which has recently been created for use in arid countries.

During 1990 alone, the U.S. Department of Agriculture (USDA) approved over 100 test plantings of crops that had been genetically altered to give them such traits as pest resistance and tolerance to weed killers.

It might be worth making a class collection of new examples of genetic engineering and posting them as they are noted in newspapers and magazines. Who benefits from the development? Who, if anyone, is hurt? How are society or people's lives altered? What hazards are associated with each example? The preceding three examples may serve as a starter.

(See "The Geopolitics of Genes," p. 18; "Citizens' Responsibility to Society," p. 119; and "Animal Rights," p. 14.)

Think about this . . . *The last remaining smallpox viruses in the world exist in only two high-security laboratories in Moscow, U.S.S.R., and Atlanta, Georgia. Should the viruses be destroyed?*

Do you say "thank you" to the recorded telephone voice answering your request for a movie timetable, the weather, directory assistance, etc.?

Issue 8

Animal Rights

Do all animals have an inalienable right to life?

If they do, consider the millions of rats, mice, rabbits, dogs, cats, and other animals sacrificed yearly in the U.S. in support of research of many kinds. Without their sacrifice, the development of thousands of new vaccines, surgical procedures, and drug therapies, and tests of potential carcinogens and new pharmaceuticals would be restricted or stopped.

In recent years, computer models and cell cultures have greatly reduced the need for live animals for many research purposes, yet many crippling and lethal human diseases remain that pose problems so complex that only live animals offer hopes for clues.

Chimpanzees are the only adequate animals on which to test new vaccines against human viral diseases. Only infant chimpanzees can be imported successfully, and their mothers are usually killed during capture. The demand for chimpanzees is so great that the species is now depleted and in danger of extinction, yet no other species can serve the purpose.

Is the conquest of human disease worth the destruction of this (or any other) species?

Unlike chimpanzees, baboons are monkeys, not apes, and are, therefore, less closely related to humans. But with humans they share similar organ architecture and several blood types—A, B, and AB—so they are important as a source of organ transplants requiring compatible blood types. A colony of baboons is being raised at a research facility near San Antonio, Texas. It was the heart of one of these baboons that was transplanted into Baby Fae, a human baby, in 1984.

Organizations opposed to such use of animals are proliferating. Legislation limiting such use has been passed already in many states and localities and is even pending in Congress. In addition, the treatment of research animals (except rats and mice) is federally regulated by the 1966 Animal Welfare

Act, amended in 1985. The trend is not so much concerned with the humane treatment of laboratory animals as it is with *animal rights,* asserting that all sentient beings have a moral right to life and, therefore, deserve legal protection from harm regardless of possible benefits for human beings.

In a survey by the *Boston Globe* newspaper, only a third of the respondents answered in the affirmative when asked, "Do you believe scientists should be allowed to experiment with live animals?"

In a poll by a well-known fashion magazine, 59 percent of respondents said they would be willing to use a drug that had not been tested on animals, even if it might not be safe.

How would you have answered those two questions?

Are there moral differences between using animals for food or clothing and using them for research or for testing consumer products? What about research on a fatal human disease?

The only survivors of the year 3000

Many people say zoos are cruel and should be banned. Others say zoos are a crucial last-ditch refuge for the rescue of vanishing species. What do you think?

Americans love violent spectator sports. How do you feel about the Spaniards' and Mexicans' attachment to bullfights? Would you like to watch one?

How do you feel about raising wild animals as pets? Suppose you come upon a nest of tiny, helpless baby squirrels (birds, rabbits, mice, etc.). Their mother is nowhere to be seen; they look abandoned. What should you do? In most cases, you should do nothing. It is usually impossible to domesticate a wild animal or to rear it and return it to the wild successfully. Moreover, it is illegal for individuals to transport wild animals across state lines anywhere in the U.S.—even baby birds or mice. Do you think this is a good law?

If you feel strongly about the protection of animal rights, how do you justify killing cattle, chickens, pigs, fish, and so forth for food? If you are a convinced vegetarian, how do you justify the right of animals to kill plants by eating *them?*

Does a beautiful tree have rights? What about our much-abused planet, Earth? The Endangered Species Act affords legal rights to *places* by protecting the habitats of species threatened by extinction. The Wilderness Act, passed 25 years ago, also preserves wilderness areas from permanent intrusion by human beings.

Think about this . . . *Because of their knowledge, do some scientists have a power that makes them dangerous?*

Issue 9

Vitamin C and History

Certainly diet has consequences for people, not only today but even five centuries ago. As an example, consider King Henry VIII of England.

"His appetite for food and for women is legendary, but despite his royal girth, Henry VIII may have died from malnutrition. According to historian Susan Maclean Kybett, it was not syphilis, as once commonly believed, but a chronic lack of vitamin C that killed the King in 1547 at age 55." ("Henry VIII—Malnourished?", *Time,* Vol. 134, Sept. 11, 1989, p. 51.)

Scurvy, caused by a lack of vitamin C in the diet, was common in Henry's era. Fruits and vegetables, primary sources of vitamin C, were scarce. Royalty considered them unfit to eat. Henry was plagued with frequent colds, constipation, a bloated body, a collapsed nose, bad breath, leg ulcers, and mood swings, which are all symptoms of scurvy. Kybett, the article says, suggests that Henry's vitamin C deficiency also affected his decisions to marry six times, to have two wives beheaded, and to break with Rome to found the Church of England.

If vitamin C tablets had existed in Henry's time, would just a few bottles of them in his diet changed the path of history?

Might diet be having any effect on events or on culture in parts of America today? How do you know your diet is adequate?

Think about this . . . *Should the deep ocean be considered an option for hazardous (radioactive or toxic) waste disposal? What about shooting the waste into space?*

Issue 10

The Geopolitics of Genes

Who actually owns the genes of the hybrid tomato plants, corn, or potatoes on which so much of your diet depends?

The agriculture of advanced industrialized nations of North America and northern Europe are based almost entirely on plants acquired decades or centuries ago from what are now developing nations of the Third World. These plants include corn, soybeans, wheat, potatoes, alfalfa, barley, sorghum, tomatoes, cotton, tobacco, and flax.

Until recently, such plant material and its germ plasm has been considered the "common heritage" of humanity, free for the taking, worldwide. Over the years, however, plants gathered and transported as "common heritage" have become worth billions of dollars to the developed nations to which they were taken. As one example, the genes from Turkish barley that gave it resistance to yellow dwarf disease is worth $150 million annually to U.S. farmers.

In contrast, the genetically perfected plant varieties bred by the commercial seed companies have been considered to be private property and, therefore, an article of commerce for which payment must be made—even by developing countries that are impoverished and dependent on them. After all, the plant breeders in the industrialized nations have given the new strains economic value by reason of their investment and labor.

Developing countries now question this difference, arguing that if the raw materials they provide are part of the world's common heritage, the genetically perfected seeds and crops developed from them must be too. Settling the debate will require an international forum and the solution to an ethical problem that has no precedent. The matter is given particular urgency by widespread awareness that more and more plant species are being extinguished worldwide by the destruction of their habitat and that the plant world's genetic base consequently is being eroded steadily.

What do you think is a fair solution?

Issue 11

Photosynthesis: Greenhouse Effect

Trees counteract the greenhouse effect by absorbing carbon dioxide from the atmosphere. If we plant enough trees, they will ease the present threat of global warming caused by the buildup of carbon dioxide in the air.

Trees, crops, grass, and other green plants absorb carbon dioxide from the air and convert it to plant material (cellulose, starch, sugars, etc.) and oxygen by a process called *photosynthesis*. The reaction is energized by sunlight and catalyzed by green chlorophyll.

$$(6CO_2 + 6H_2O + light \rightarrow C_6H_{12}O_6 + 6O_2).$$

Carbon dioxide is the gas chiefly responsible for the greenhouse effect. (See "Absorption of Radiation: Greenhouse Effect," p. 92.) Its removal from the atmosphere by the photosynthesis of plants is being slowed by the massive destruction of forests and the desertification of agricultural land worldwide. Instead, its concentration in the atmosphere is being increased steadily by the burning of such fossil fuels as coal, oil, and gasoline. (See "Destruction of Tropical Forests," p. 33.)

To reduce the greenhouse effect, we clearly must reduce the concentration of carbon dioxide in the air, which requires that we decrease fossil fuel consumption and increase the effectiveness of photosynthesis.

These fundamental facts of nature have convinced a number of experts that the widespread planting of trees and the conservation of existing forests is one of the surest, easiest, and least expensive ways to slow or even reverse the buildup of carbon dioxide in the air.

In one of the first large-scale actions in this direction, the American Forestry Association, a citizens' conservation organization, has undertaken a national campaign aimed at planting 100 million new trees in American cities and towns by 1992. The Environmental Protection Association presently is considering the feasibility of planting 400 million trees. These trees could reforest 20 percent of the U.S. highway corridors and, planted strategically around buildings, would not only absorb carbon dioxide but provide shade to help reduce energy use in hot weather. The EPA estimates that the total mass of American forests could be increased by 60 percent in this way.

Here is something you can do to help in a practical way.

It makes sense to develop large-scale tree planting programs on degraded land in the tropics, some of it denuded and abandoned after traditional slash-and-burn. A reforested acre in your hometown in the U.S. is just as effective as a reforested acre in Brazil or Indonesia in helping promote photosynthesis worldwide and counteracting the greenhouse effect. Thus, by planting trees, even in your own backyard, you can genuinely address a world-class environmental problem without having to travel overseas.

Over 100,000 children and adults in 42 states have planted and cultivated many thousands of trees for fruit or shade through the agency Trees for Life. Each tree removes an average of 6 kilograms (13 pounds) of carbon dioxide every year, converting it to oxygen, as shown in the above equation.

Here are more things you can do to fight global warming:

— Recycle newspapers. An entire printing of the Sunday *New York Times* newspaper consumes an average of 75,000 trees; preserving them by recycling could remove nearly half a million kilograms (about 1 million pounds) of carbon dioxide annually.

— Switch to natural gas. Burning coal and oil for energy are far greater culprits in global warming than is natural gas.

— Insulate your space. Each year, as much energy leaks through American windows as flows through the Alaskan oil pipeline.

— Ride the bus or subway whenever possible. About one-third of the global warming caused by the U.S. stems from transportation.

— Get better gas mileage. The average car driven in the U.S. emits nearly 54 metric tons (60 tons) of carbon dioxide during its lifetime. Fuel-efficient cars available today emit less than half that amount.

— Put the heat on politicians. There are scores of actions our elected leaders should take at the national, state, and local levels.

Does it follow that because you are so small and the problem is so big that your effort will be worthless?

(See "Trees vs. Desert: A Project," p. 22.)

Issue 12

National Parks

In Tanzania, Africa, Masai cattle raisers annually burn the old withered grass so fresh grass will sprout better in the spring rains to feed their herds. The fires also destroy bushes and woodlands, increasing the grassland area. As a consequence, the population of grass-eating wild animals in the adjoining Serengeti National Park has expanded enormously. The 100,000 wildebeests there in 1960 soared to 2 million in 1980. Gazelles and zebra populations have also exploded; in a drought, overpopulation results in mass starvation. One way to prevent this is to harvest a fraction of the population for meat and hides and for export. This means the regular, systematic slaughter of animals in one of Africa's most beautiful and unspoiled national parks. Should this be done?

If oil were found beneath one of *our* national parks, should it be drilled out? What about Alaska's Arctic National Wildlife Refuge, 7.7 million hectares (19 million acres) of fragile especially rich unspoiled coastal plain?

"Honey, it's oil! Quick . . . call our real estate agent in Beverly Hills!"

Issue 13

Trees vs. Desert: A Project

Your school parking lot and possibly the playground may be barren, dry, sunbaked, and lifeless—ecologically a desert. Like real deserts in Third World countries, they are not apt to support useful forms of life.

Perhaps they may be made more attractive and even productive by the use of a technique now being used abroad to combat desertification—the planting of trees.

Some species of trees are especially suited to dry, barren areas where nothing else now grows, like your school parking lot or playground. Depending on the local conditions and needs, trees may be chosen to provide windbreaks, shade, food for people, erosion and dust control, fodder for animals, fuel, industrial raw materials, and pollen and nectar for honeybees. Some grow as much as 3 meters (10 feet) a year! Your school may reap all these benefits plus a saving of energy—if there is sufficient shade, drivers do not have to turn on air conditioners as they get into overheated parked cars. Moreover, there is almost always a big aesthetic improvement.

Perhaps the most valuable educational benefit of launching a tree-planting project will be the experience of dealing with all the necessary bureaucratic problems typical of efforts to convert sound scientific principles into practical reality! Here in the microcosm of your school community is a sample of what is happening in international politics and finance as developed nations help impoverished developing countries in their struggle to hold back encroaching deserts.

If you like the idea but think "nothing can grow" in the school parking lot (asphalt removed!) or playground, remember that's true of deserts too. Desert-tested varieties of carob, honey locust, prosopis, acacia, eucalyptus, and casuarina, among others, may be able to grow and stabilize infertile soil in your part of the country. Deserts, incidentally, cover 23 percent of the earth's land surface and are increasing at the rate of 64,750 square kilometers (25,000 square miles) every year.

(See "Destruction of Tropical Forests," p. 33 and "Photosynthesis: Greenhouse Effect," p. 19.)

__Think about this__ . . . Should free needles be issued to drug addicts in order to combat the spread of AIDS?

Is there a moral difference between killing animals (mammals) in the course of (__a__) testing consumer products, (__b__) medical research, and (__c__) producing food and clothing?

Should cities and states confronted with a shortage of water pass legislation restricting community growth?

Issue 14

World Population Explosion

If the present human birthrate continues, by A. D. 2600 there will be only 0.84 square meters (1 square yard) of dry land surface available for each inhabitant—hardly more than the size of a telephone booth. Today, at the rate of 163 births per minute, a baby is born somewhere in the world every ⅓ second, and the world's population is growing at the net rate of 235,000 people per day (birthrate minus death rate). It has doubled since 1952 and is increasing at 1.7 percent per year. (See "Exponential Growth," p. 112.)

The world population, which stood at 1.5 billion in 1900, was 5.2 billion in mid-1989 and is expected to grow to 6.2 billion in the year 2000 and to 8.5 billion in 2025. It may reach 10 billion before it levels off toward the end of the next century, with nearly 9 out of 10 persons living in developing countries (U.N. median projections).

No government or academic expert has the faintest idea how to provide adequate food, housing, health care, education, and gainful employment to such exploding numbers of people as they crowd into such mega-cities as Mexico City, Calcutta, and Cairo with their inadequate housing, lack of health facilities, and overcrowded schools.

Moreover, the growing numbers of desperate poor implied in these figures will accelerate the ferocious assault on the environment already under way in Africa, Asia, and Latin America as they overgraze grasslands, cut and burn forests, and over-plow croplands in a desperate effort to produce more food.

Should the size or growth rate of the world's human population be curtailed to protect the global environment? If so, how? If not, what?

Has society the right to tell individuals (including you) how many children they can have? If your answer is "no," may population control be achieved by an appeal to conscience? What form of coercion or appeal *would* you respond to?

(See "Tragedy of the Commons," p. 110 and "Exponential Growth," p. 112.)

Think about this . . . *Any chemist who perfects a blackfly and mosquito repellent could single-handedly alter the lifestyle and economy of large parts of rural Maine. For years the biting insects have protected some of the most beautiful and underpopulated areas of the state from floods of vacationers and home builders. A single effective use of chemistry could suddenly make it possible for many people to move in. Is this a change to be welcomed?*

Issue 15

Population Paradox

Most people consider overpopulation to be one of the most serious threats to the quality of life and to world peace. Do you agree? These things make overpopulation even greater and hence *increase* the population problem: motherhood, medicine, public health, peace, law and order, scientific agriculture, accident prevention, and clean air.

These things *reduce* the population problem: disease, abortion, war, murder, famine, accidents, family planning, and sexual abstinence.

How do you resolve this paradox?

Think about this . . . Given time and money, can science eventually solve such social problems as crime, hunger, and mental illness?

Because it unquestionably saves lives, shouldn't the original 55 mi/hr speed limit on interstate highways be retained and enforced more rigidly?

Issue 16

U.S. Birthrate

Every 8.5 seconds a new American is born. It is a disarming little thing but begins to scream loudly in a voice that will be heard for 70 years. It is screaming for 23.6 metric tons (26 million tons) of water, 79,500 liters (21,000 gallons) of gasoline, 4600 kilograms (10,150 pounds) of meat, 12,700 kilograms (28,000 pounds) of milk and cream, 4080 kilograms (9,000 pounds) of wheat, and great storehouses of other foods and drinks. These are its lifetime demands on country and economy.

The *birthrate* in the U.S. is now at about replacement level. That is, each couple produces, on average, two children who will in turn survive and produce two children—no more. Yet, the population keeps growing, quite apart from immigration. Why? (Answer: Because of the population "hump" of parents at the child-bearing age). The net growth rate of the population is 1 additional American every 21 seconds. The 1990 population was close to 245 million. What effect is this near-zero birth rate having on:

— the public school system?
— U.S. taxpayers' contributions to Social Security?
— the average age of U.S. citizens?

Suggest a program whereby the U.S. might achieve and maintain *zero* population growth. Would your program be politically practical? Who would pay for it? Is the U.S. immune from the effects—and prospects—of the worldwide population explosion? What will be some of its effects on you? What should be done, if anything? (See "World Population Explosion," p. 24.)

Think about this . . . *Should science courses be required for obtaining high school (or college) diplomas? Why?*

Issue 17

Pesticide Pollution

Residents in expensive new housing developments on the west side of Denver, Colorado, were severely troubled by mosquitoes (1984). Even heavy spraying of pesticides did not eliminate the pests; it only killed off harmless wildlife and began to contaminate the groundwater. Realistically, what should the residents do? Go on spraying? Stop spraying and suffer? Sell (if they can) and move out? Or . . .?

In Suffolk County, Long Island, the leading farm county in New York, chemicals are losing the battle against the Colorado potato beetle. The beetle has acquired resistance to all major pesticides registered for use on potatoes. Growers spray up to 10 times per season, and pest control costs have climbed as high as $700 per hectare. Meanwhile, heavy application of pesticides has caused extensive contamination of groundwater, the region's sole source of drinking water.

Here in the U.S., tests show that more than 25 percent of Iowans use drinking water contaminated with pesticides.

Might the science of biotechnology (genetic engineering) offer a remedy to either of these problems?

In developing countries, pesticides poison people as well as insects. The U.S. annually exports 225 million kilograms (500 million pounds) of pesticides that are banned, restricted, or not licensed for use in the U.S. Overseas, typically in hot Third World countries, the chemicals are put to work in the constant uphill fight against insects, plant diseases, and weeds that threaten the production of food. But in many countries it is so hot that the standard protective clothing designed for use in temperate climates is unbearable. (Try working in an airtight rubber suit in the sun at 37.7°C (100°F) for hours at a time!) Many of the workers are untrained and cannot read or understand the instructions. (Rural Brazil, a major user of pesticides, is 40 percent illiterate.) Is it surprising, then, that people use the empty pesticide containers to store drinking water and food?

Eighty-five developing countries urgently need to use the chemicals, yet at the same time they suffer between 10,000 and 40,000 annual deaths from the chemicals and as many as a million cases of poisoning. A concerned U.S. and the European Community are attempting to address the problem. If you were an official in the United Nations, where much of the negotiation takes place, how would you recommend that these two conflicting needs (use vs. poisoning) be reconciled?

Ironically, Americans remain exposed to some of the poisonous chemicals banned from use in the U.S.: we export them to developing countries for use in growing food that we then import and eat, completing a "circle of poison."

(See "Hazardous Chemicals, Dioxin, and Molecular Formulas," p. 54 and "Patenting Plants and Animals," p. 11.)

Issue 18

Medfly

Hundreds of millions of dollars have been spent by the state of California on efforts to eliminate the Mediterranean fruit fly. Yet entomologists insist that the massive use of quarantine, trapping, stripping trees of their infected fruit, release of sterilized flies, and spraying with malathion are all demonstrably flawed in one way or another and cannot succeed. These experts say that eventually, because of the insect's biology, it will disappear of its own accord.

Should the state officials heed the fruit industry, which understandably clamors for some kind of action? Or should they risk their political careers by heeding the entomologists, who should know more about the insects than anyone else, and do nothing? Or should they heed the rising protests of a restive public suspicious of adverse health effects from repeated spraying of malathion on residential districts? If you were a responsible politician, what action would you take?

"Just fly real slow and act natural."

Issue 19

Destruction of Species

If you could destroy all the rattlesnakes in the U.S., would you do so? What about mosquitoes? Rats and mice? How would your life be different if there were no cats and dogs? If all the birds were to disappear?

Has humankind an obligation to preserve the species of plants and animals we find on the earth? Or should they exist only as long as we find it convenient to have them around?

The California condor is an ugly bird, and only a few survive, so who cares if they become extinct? Warthogs, bats, and octopuses are considered by many people to be ugly; the leopard is considered to be beautiful. Should beauty determine our sentiment toward conservation?

Wildlife officials are spending taxpayers' money to bring back from near extinction the grizzly bear, timber wolf, alligator, bald eagle, and Florida panther, all carnivores. Does this effort make sense?

Worldwide, the destruction of habitats is driving to extinction at least ten plant and animal species every day. Of the earth's 5 million or more species, more than 1 million could be lost by the end of the century. We are in the midst of an extinction of plants and animals a 1000 times more rapid than the pace of evolution since multicellular life appeared about 600 million years ago. It is well documented that it is the result of human activity. Should we care? Why?

Losses are especially severe in tropical forests. They are enormous gene pools, home to at least 2 million plant and animal species—50 to 80 percent of the planet's species—many of which have not yet even been identified. The genetic material being destroyed (some 4000 species a year by one conservative estimate) may contain secrets for improving food crops or fighting disease or regenerating soil or as sources of oil or new fibers—and it is being lost forever.

As a result, zoos are acquiring a new importance. No longer merely museum collections of animals, many are becoming rescue centers where nearly extinct species can be bred and protected.

In 1960 a person with leukemia had 1 chance in 5 of survival. In 1980 the outlook was 4 in 5. This change stems in part from the discovery that a tropical forest plant, the rosy periwinkle, contains alkaloidal material (vincristine and vinblastin) with powerful antileukemia properties. Of 32 such commonly used chemotherapy drugs, 28 are plant products discovered by chance.

Antibiotics resulted from observations of a mold growing on a cantaloupe. Smallpox elimination resulted from observations of the effect of cowpox virus on the health of milkmaids. The peregrine falcon's fragile eggs warned us of excessive DDT in our environment. The Devil's Hole pupfish warned of a falling water table in Nevada. Hamsters (medical science) and rubber trees (technology) have had immense effects on our lives.

When we buy medication, there is a 50 percent chance it derives from materials of natural origin. Scientists have screened for usefulness only 1 percent of the earth's plant and animal species so far. How effective is the Endangered Species Act? Why are mining and utility companies often anxious to eliminate many of its most important provisions? In the U.S. alone, 500 species of plants and animals are listed as endangered. Approximately one-third are losing the battle for survival. Lack of money to protect their habitats is part of the reason.

Endangered plants have received less attention than endangered animals. In the U.S., botanists have estimated that about 3000 plant species, more than 10 percent of the estimated 25,000 species of native U.S. plants, are in some danger of extinction. Worldwide at least 50,000 plant species have edible parts, but humans rely heavily on only about 20 of them. (For information on endangered plants in the U.S., write to the Center for Plant Conservation, Inc., Arnold Arboretum, Arborway, Jamaica, MA 02130.)

A project: Identify endangered plants and animals in your state or area. What are the threats to their survival and what is being done? Can you help?

Many of the genetic resources on which U.S. agriculture is based are from overseas. It is important to prevent the loss of wild varieties of grain and ancient lines of livestock that could be valuable someday for breeding. Couldn't Congress be encouraged to tie the conservation of genetic resources to international aid?

The search for extraterrestrial life was the main motivation for the 1976 *Viking* mission to Mars. It cost NASA about $1 billion, and was in vain. How much is it worth to search, with *guaranteed* success, for species that are vanishing from the earth? (See "Destruction of Tropical Forests," p. 33, and "Worthless Species?" p. 35.)

Issue 20

Destruction of Tropical Forests

Worldwide, an area of tropical forests the size of Pennsylvania (117,000 square kilometers or 45,000 square miles) is being cut down every year (1989). That is almost 13 square kilometers (5 square miles) every *hour,* day and night. The Amazonian forests of Brazil, largest remaining forests on the earth, were being cut (in 1987) at the rate of 310 square kilometers (120 square miles) every day. The reasons: fuel, timber, cattle ranching, and space for exploding populations to grow food.

Forest clearing removes most of the nutrients from the soil, destroying its fertility to such an extent that large areas of pasture and cropland that replaced tropical forest have, in Brazil for example, already become useless wasteland and been abandoned.

The effects include (a) major but unpredictable changes in world weather patterns, including enhancement of the greenhouse effect, which in turn may affect agriculture and create unpredictable changes in the world's food supply—and your food bill, and (b) the destruction of thousands of species of plants and animals.

It is worth noting that *increasing* reforestation worldwide between now and A.D. 2000 by an area just twice the size of Texas would return the world's supply of wood for fuel and industry to a sustainable level. It also would capture and store considerable carbon, reducing the rate of carbon dioxide buildup and global greenhouse warming.

Should anyone be doing anything about this situation? If so, what? Perhaps a social studies teacher has some suggestions.

Bear in mind that in many poor tropical countries the forest destruction is most rapid. Setting aside large areas of forest for conservation takes it out of the economy of the poorest and least fertile agricultural region where the poorest people must try to make a living. How can a rain forest be used productively without destroying it?

(See "Trees vs. Desert: A Project," p. 22, "Tragedy of the Commons," p. 110, and "Destruction of Species," p. 31.)

Question: Do you think it is right to buy imported products made from such tropical woods as rosewood, mahogany, teak, or cocabola? On one hand, you help the economy of a poor developing country today; on the other hand, your purchase contributes to that country's long-term impoverishment. Do you approve of the consumer boycott of Burger King which, in the summer of 1987, forced the company to stop importing beef from Costa Rica, where cattle ranches were expanding on land being cut from rain forest?

Think about this . . . *If a fetus has rights, isn't a crack- or cocaine-using pregnant mother guilty of drug pushing?*

Issue 21

Worthless Species?

Thanks to biotechnology, we are learning how to create new forms of life tailored to specific purposes. (See "Patenting Plants and Animals," p. 11.) If it is all right to create desirable new species, is it all right to destroy undesirable or useless species? After all, we already are spending immense amounts of money to destroy or control insects that kill crops and spread disease. If you had the power to do so, would you destroy:

— the AIDS virus?
— the tuberculosis bacterium (which kills 3 million people worldwide per year)?
— mosquitoes (they spread malaria) and tsetse flies (they spread sleeping sickness)?
— cockroaches, poisonous snakes, or alligators?
— weeds?

Where do you draw the line? Why there?

Without affecting humankind, species of plants and animals have come and gone since life began on the earth. The dinosaurs are an example. Why should we be concerned about extinctions occurring today?

See "Destruction of Tropical Forests," p. 33, and "Destruction of Species," p. 31.)

> ***Think about this . . .*** *To reduce roadside litter, should state laws be passed that ban the use of throwaway containers for beverages such as beer and soft drinks— or require a refundable deposit?*

Issue 22

Human Gene Manipulation

As of 1989, human gene manipulation could only be performed on tissue cells (such as bone marrow cells), so the altered cells died with the patient. The day may come, however, when it will be possible to alter reproductive cells, in which the altered genes—good or bad—will be transmitted from one generation to the next. Here lies cause for serious concern. Medical scientists, like doctors everywhere, have a responsibility to protect the public against diseases of all kinds. When it becomes possible to alter human reproductive cells to prevent defective genes from passing from parents to children, it may become criminal *not* to do so.

Many ethicists see nothing wrong with experiments that genetically alter microorganisms or plants or even large animals. They do, however, express doubts about experiments to modify the genes of human beings. They foresee a series of progressively more difficult ethical questions. For example:

— Should alteration of the genes of an individual with a previously incurable genetic disease such as Huntington's disease or Tay-Sachs disease be allowed?

— Should alteration in that individual's reproductive cells be allowed so that his or her children will not inherit the disease? If asthma proves to be a genetic disease, should that be cured genetically? What about baldness?

— If it becomes possible to do so, should alteration in any *healthy* individual's reproductive cells to "improve" his or her children by making them taller, stronger, or smarter be allowed?

— What if people want to improve genes that are not defective but merely mediocre? Could genetic engineering become the cosmetic surgery of the next century?

In June 1983, a group of about 40 religious leaders issued a statement calling for a complete ban on human genetic engineering, or gene transplants, arguing that "humans should not assume the prerogatives of the Creator," and that "no individual, group, or institution can legitimately claim the right or authority to make such decisions on behalf of the rest of the species."

On the other hand, many people contend that the chance to reduce suffering or enhance human well-being far outweighs the risks of genetic tampering. Who do you think is right?

Just before and during World War II, the Nazis used political action and physical force to promote what they considered to be "pure Aryan" German citizens. They set about destroying all the Jews they could (about 6,000,000). Does this fact alter your answers to the question above?

(See "Animal Rights," p. 14 and "Patenting Plants and Animals," p. 11.)

Think about this . . . The U.S. Forest Service owns half the softwood timber in the U.S. Should there be a connection between their sale of trees for harvesting and the nationwide recycling of waste paper? If so, what?

Rich coal beds underlie some of the most fertile wheat-growing land in the American West. As oil grows increasingly scarce and expensive, should the coal beds (which lie near the surface) be strip-mined?

Issue 23

Fetal Medical Examination

Should parents be able to learn—or choose—the sex of their unborn child?

A number of techniques now provide increasingly reliable identification of the sex of an unborn child, as well as a diagnosis of physical deformities and such genetic diseases as sickle-cell anemia, hypothyroidism, and Huntington's chorea. The techniques include biochemical and hormonal tests, ultrasound, amniocentesis, chorionic villi biopsy, and even fetoscopy. But the use of such procedures raises such concerns as:

— the fetus's right to privacy—a right that may invoke the U.S. Constitution.
— when does humanness begin?
— the right of the fetus to life.
— the protection of the defenseless.
— the rights of society with respect to the welfare of its citizenry.

With the dramatic rise in teenage pregnancy, these are more than theoretical issues. For example, should a doctor perform an amniocentesis for an expectant mother to determine the sex of her unborn child if the mother, who is strenuously opposed to its being a boy, will have an abortion if it is to be a boy?

Should prenatal diagnosis of sex be prohibited by law? About 15,000 amniocentesis are done annually in the U.S. (1987).

In desperately overpopulated China (one-fourth of the entire world's population), parents are penalized by the government if they have more than two children. Yet traditionally in both China and India, where there is no such thing as life insurance as we know it, parents are supported in their old age by their sons, not by their daughters. What would be the effect in both China and India of widespread, easily accessible, amniocentesis?

If a fetus has rights, isn't its cocaine-using mother guilty of drug pushing? If a woman has found that her child has a major but not life-threatening defect and wishes to abort, should her doing so be viewed as murder?

Think about this . . . *Are people in our society who understand science and technology better off than those who do not?*

"Doctor, is it too late? We've changed our minds!
We've decided we want a girl instead."

Issue 24

AIDS and Society's Responsibility

Both for doctors and for society, AIDS raises problems unlike those of any other disease.

For society the problems turn on the fact that most AIDS patients cannot afford their long period (6 months to 2 years) of terminal care. Hence, insurance companies try to screen out applicants for insurance and to deny policies to applicants who may be carrying the disease. To the degree they are unsuccessful, the company's funds for terminal care are being consumed at the expense of other policy holders, whose premiums must be increased. Hospital budgets are being strained and broken to the detriment of patients whose illnesses are fundamentally more manageable.

Someday (if not now) you probably will be an applicant for medical insurance. How can insurance companies and hospitals be fair to you and at the same time do justice to the needs of AIDS sufferers?

For doctors, one of the problems posed by AIDS was expressed by Dr. C. Everett Koop when he was the U.S. Surgeon General, speaking to the President's AIDS Commission (*The New York Times,* Sept. 13, 1987):

"Health care in this country has always been predicated on the assumption that somehow everyone will be cared for and that no one will be turned away . . . that care will not be abandoned for the sick and the disabled, whoever they are. Hence, the reports of a few physicians and others withholding care from persons with AIDS are extremely serious, in my opinion. Such conduct threatens the very fabric of health care in this country."

Remembering that however unlikely it may be for a physician or a dentist to acquire AIDS in treating a patient, and remembering that the disease is invariably fatal, what are the responsibilities of a doctor who may have a family of his own to consider as well as an AIDS patient? Just such a case occurred in Florida in the autumn of 1990 when a woman in Florida was given a verified case of AIDS by an AIDS-infected dentist.

Issue 25

Organ Transplants

Many parts of your body can be replaced by artificial parts (prostheses). Plastic hips, hearing aids, pacemakers for your heart, wigs, dental structures, and artificial arms and legs are only a few of the more familiar ones.

Many more parts of your body can be replaced by transplanting living, healthy organs from someone else. Cornea and kidney transplants, skin grafts, and blood transfusions are common. Entire hearts are now being transplanted from animals and from accident victims into people with terminal heart disorders. The need for transplant organs now far exceeds the number of donors becoming available. This situation raises tough questions.

As a first example, consider the problems raised by the scarcity of organs. Given two desperately needy patients, both about to die, how do you decide which one is to be saved by the one available heart from an accident victim? Would you be willing to donate one of your two kidneys to save the life of a stranger? Of a family member?

Second, fresh transplant organs typically are acquired from the body of someone who has just died or been killed in an accident. Speed is important in performing the transplant, but how do you know the donor is truly dead? Brain waves? Heartbeat? Breathing? Who has the right to grant the use of a donor's organs? Relatives (often unavailable)? The surgeon (a heavy responsibility for one person who is a stranger)? A committee (hard to assemble quickly)? Although delay may cost the life of the recipient, haste also may cost the life and rights of the donor.

Third, in France a law decrees that at death all the organs and tissues of a person's body may be used for transplant purposes unless he or she objected in writing before the death. This measure makes available a copious supply of organs for transplant purposes—hearts, lungs, livers, kidneys, corneas—and undoubtedly will save numerous lives. Do you think a government has the right to decide the use of your body at death, even for humanitarian purposes? On the other hand, isn't it selfish, if not almost criminal, to refuse the use of your dead body for humanitarian purposes?

Fourth, it is becoming practical to implant an entire artificial heart into an ailing human (Barney Clark, 1983; William Schroeder, 1984). About 50,000 people a year stand to gain added years of life from the operation, which costs $100,000 (1988). But that same total investment of $5 billion would save many more than 50,000 lives if it were invested in measures to prevent heart disease in the first place. (See "Tobacco," p. 8.)

In 1983, 172 transplants of natural hearts actually were performed in the U.S. As many as 5000 people could have benefited from the operation. The available organs often go to those who can pay. Should people be allowed to die for lack of money to pay for medical treatment?

A new and practical use for a transplant: Perhaps in jest, it has been proposed that people liable to be kidnapped (e.g., important business people or public figures) have tiny radio transmitters implanted somewhere in their bodies so that they and their kidnappers can be tracked!

Think about this . . . Putting fluorine into the public drinking water supply reduces cavities in your teeth, but it is medication without your consent. Should it, therefore, be stopped?

Does a garden fertilized with synthetic chemical fertilizer grow as good a crop as one fertilized with natural plant and animal material?

Issue 26

Medical Tests for Health Insurability

Should a test that may predict a serious illness in the future be a condition for getting health insurance or a job?

For example, a positive test for HIV (human immunodeficiency virus) in a person's blood means a 25 to 50 percent chance of his or her coming down with AIDS within 5 to 10 years of infection, followed by certain death. That test result, in turn, may make an otherwise healthy person uninsurable.

Insurance companies in some states are demanding proof that an applicant is not at risk for AIDS. Since the number of people at risk for AIDS is rising to epidemic proportions, so also is the number of people who are unable to obtain health or life insurance and are, therefore, destined to become wards of the state when they become ill. If blood tests are legally prohibited as a condition for obtaining health insurance, insurance companies will be unable to screen out people who are at particularly high risk. As a result, the increasing number of AIDS victims will raise the price of policies for *all* insured individuals. In effect, the enormous cost of the AIDS epidemic will be increasingly borne by the uninfected public.

Should blood tests be made mandatory for insurance applicants? If you say "No," who should pay the costs of hospitalizing people who later come down with AIDS? If you say "Yes," you should recognize that a tiny minority of people who do not have HIV in their blood nonetheless test positive for it. To exclude them from insurance or jobs clearly would work a great injustice to these very few.

AIDS victims are not the only people at risk. There has been progress in mapping human genes. In the near future, patterns of genes associated with a predisposition to heart attacks, diabetes, Alzheimer's disease, and certain cancers may be identified quickly and simply. When perfected, these tests will make it possible to identify a person even in the womb who may later in life face a long and costly battle for health—and in the eyes of health and life insurance companies be uninsurable. As a result, healthy people pay the medical expenses of the unfortunate people who are sick. Is this right? If not, what is your solution?

Suppose it becomes legal for companies to withhold insurance from people whose life and health are found to be at risk at some indefinite time in the future. Is it probable that prospective employers will make negative blood tests a condition of employment? Why should a company hire and go to the expense of training someone who will probably (or certainly) come down with AIDS or the consequences of some congenital gene deficiency in a few years?

Is it fair to forbid employers to make a blood test a condition for getting a job? Is it fair to the AIDS carrier to be denied a job while he or she is still physically fit and otherwise qualified?

Should blood tests as a condition for employment or insurance coverage be legal? Are they an infringement on an individual's right to privacy?

(See "Fetal Medical Examination," p. 38.)

"I'm sorry, Mr. Superman, but this allergic reaction of yours to kryptonite excludes you from medical coverage."

Chemistry

As soon as questions of will or decision or reason or choice of action arise, human science is at a loss.

— Noam Chomsky

Issue 27

Law of Conservation of Matter

Where is all the trash and waste material you produced during the past month?

The law of conservation of matter is one of the cornerstones of chemistry. Just about all of the atoms you are made of have been around since the formation of the earth—and earlier—and have been used over and over again by living organisms down through history. And they will persist into the indefinite future.

The atoms of wastes (if they are not radioactive) are also indestructible. While some waste compounds decompose into simpler compounds (e.g., sewage and kitchen garbage), other compounds often do not. Whenever it rains, our lakes, rivers, and shorelines are dosed with oil, lead, and bacteria from city streets, sediments and pesticides from eroding farm lands, soil from construction sites, and acids from mines.

An example of mine acids is the "acid mine drainage" associated with coal deposits. It contaminates streams in coal-mining areas with sulfuric acid. A typical reaction is:

$$4FeS + 9O_2 + 4H_2O \rightarrow 2Fe_2O_3 + 4H_2SO_4.$$

The FeS (iron sulfide) is from leftover deposits; the O_2 (oxygen) is from the air.

Another source of waste is the rubber dust from the millions of car tires wearing down on our streets and freeways. Where does all that rubber go? "A lot of it goes up and sticks onto our windows. But only up to about the third floor. Most of it cuts out at about the seventh floor, which is one reason that more important people have higher offices. Us lesser folks use 135,000 gallons of Windex every day to remove rubber dust from our windows." (*Discover,* Nov. 1986, p. 112.)

Flushing the toilet, pouring chemicals down the sink, and throwing old tires into the river are other familiar examples of thoughtless disposal. But the law of conservation of matter says that those undesirable polluting molecules cannot be destroyed; they can only be moved somewhere else (like where?) or expensively converted into less polluting material.

Who should pay for the moving or converting?

If industrial firms cannot afford to clean up and still stay in business, jobs are lost and an entire community may decline. If taxpayers cannot afford to clean up chemical waste dumps, community health is endangered. The problems of pollution in our environment are one consequence of the law of conservation of matter.

(See: "Waste Disposal," p. 57.)

**"Yes, my career was very wearing, but now that
I've retired, I feel as if I could go on forever."**

Issue 28

Randomness: Entropy

Things fall apart; the center cannot hold.
Mere anarchy is loosed upon the land.
— W. B. YEATS,
The Second Coming

Work must be done (energy expended) to create order out of disorder. Left to themselves, orderly arrangements of matter tend to become disorderly, or random. Disorder (measured by a quantity called entropy) tends to increase. Thus:

— A pack of cards arranged in order becomes disordered when shuffled. It is extremely unlikely that shuffling would rearrange the cards into numerical order again. An effort (work) would be required to arrange them in order.
— A box of sugar cubes and an egg are orderly structures. Dropped on the floor they become disordered. (Remember Humpty Dumpty.)
— A plant or animal body is highly ordered when it is alive. Metabolic work during life maintains the order. At death, randomness ensues.
— Environmental pollutants of all kinds (chemicals, heat, solid waste) may be tightly confined in various containers (ordered). If not cared for, the containers fail, and disorder follows. Work then is required to restore order, which is usually expensive and sometimes impractical.
— It is very expensive to extract fresh water from the ocean.

For further examples, consider what happens to the orderliness of:

— soft drink and beer bottles ending up alongside highways.
— the energy of a lump of coal being dissipated into the atmosphere.
— your disorderly bedroom unless you perform work to clean it up.
— a discarded automobile rusting in the town dump.

— an abandoned building.

— PCBs and other industrial wastes disposed of in the river, in wasteland above the water table, or as flue gases up the smokestack.

— radioactive waste from the spent fuel rods of nuclear power plants.

— information transmitted by word of mouth.

Can you name other examples?

Collect and post appropriate news clippings that illustrate the increase or decrease of entropy in the environment. Note environmental degradation of all kinds, collisions, wear and obsolescence vs. repairing, building, publishing, and growth.

The General Accounting Office estimates that there are over 130,000 hazardous waste sites in the U.S. Only a few hundred have been cleaned up so far. Is this a good example of how entropy behaves? In what ways do you increase the disorder of your own world during an average day? Can you name ten of them?

Issue 29

Methane and Fermentation: Developing Countries

How well can you eat when fuel for cooking is scarce?

In the countryside of developing countries, wood is often in very short supply because most trees have been cut down. Dried cattle dung is often burned, but then it cannot be used as a badly needed fertilizer.

However, the daily dung from just three cattle is enough to generate methane gas with a calorific value equivalent to 4 or 5 liters (about a gallon) of diesel fuel. This amount is enough for an average family's domestic use. The fermentation equipment for producing and using the gas is very simple and often can be made locally in discarded oil drums.

In the reaction, cellulose and other plant matter, $n(CH_2O)$, is broken down by anaerobic bacteria into CO_2 and methane, CH_4, as follows:

$$n(CH_2O) \rightarrow \frac{n}{2}CO_2 + \frac{n}{2}CH_4$$

The methane may be burned to cook food. It may also be used for the manufacture of caustic soda, NaOH, which is of great value in many small village industries for the production of soap, dyes, and the processing of cloth. NaOH typically is bought from larger centers, where it is prepared by electrolysis, a process impractical locally. Using methane gas to heat a mixture of slaked lime and sodium carbonate in water produces NaOH in quantity.

$$Na_2CO_3 + Ca(OH)_2 \rightarrow CaCO_3 + 2NaOH$$

The raw materials are often available cheaply, and cast-off oil drums may be used to contain the reaction and to filter off the NaOH solution from the solid calcium carbonate.

Reactions like these, using simple equipment and local raw materials, were abandoned long ago in such technically advanced countries as the U.S.; they require too much labor and cannot be operated continuously. (They are "batch" processes.) But both of these factors are advantages in many developing countries.

People from developed countries who work in Third World countries overseas use elementary chemistry in a similar way to help impoverished rural areas process such basic raw materials as wood, beet sugar, plant fibers, seaweed, and animal products for either local benefit or to sell elsewhere for money. In many parts of the world, such simple village processes now create employment and produce previously unavailable materials. Each developing country has different needs and different solutions. Perhaps you will become interested in using your knowledge of chemistry in this practical way abroad someday.

For information on low-input agriculture and low-cost community resource management, write Agricenter International, 7777 Walnut Grove Rd., Box 17, Memphis, TN 38119.

(See "Energy Conversion: Developing Countries," p. 101.)

Think about this . . . No money should be spent trying to eradicate the Medfly, gypsy moth, or other unusual insect infestation. Do you agree?

Might the alteration of human reproductive cells by genetic engineering become the cosmetic surgery of the next century?

Issue 30

Ozone Catalysis:
Destruction of Our Atmosphere

Though too much ozone in ground-level smog is a health hazard, too *little* ozone in the upper atmosphere is a health hazard too!

The ozone layer in the upper atmosphere (12 to 50 kilometers up) filters out all the solar ultraviolet light shorter than 3000 Angstroms. Any significant destruction of the ozone layer will, therefore, result in an increase in cancer-causing and plant-killing ultraviolet radiation at ground level.

Ozone (O_3) is highly reactive and is destroyed by the nitrous oxide (N_2O) and carbon dioxide (CO_2) emitted by stratosphere-flying planes (for example $3N_2O + O_3 \rightarrow 6NO$) and by chlorine from Freon spray-can propellants and refrigerants, such as $CFCl_3$ and CF_2Cl_2 (called chlorofluoro-carbons, or CFCs for short). In sunlight, Freon breaks down to release chlorine, which catalyzes the reaction $2O_3 \rightarrow 3O_2$. Since each chlorine atom acts as a catalyst and therefore is not used up, it can go on wreaking havoc in the ozone layer for as long as 70 to 100 years. Continuing destruction of ozone will result in an increase in serious skin cancers and birth defects, kill plankton in the oceans, reduce the yield of many crops, and may even destroy forests and alter climate. The effects probably will take years, perhaps decades, to be fully apparent. By then, the damage will be irreversible.

There is no agreement even among experts on how much destruction of ozone results in how much increase in ground-level ultraviolet light or how destructive any given increase in ultraviolet light would be. Extensive study of the problem is urgently in order, but any agreed-on actions will take years— even decades—to have an effect. It already may be too late to avert serious consequences.

In view of these facts, what do you think should be done—and by whom? International treaties? Unilateral restraint by Freon producers? Nothing? (See "Absorption of Radiation: Greenhouse Effect," p. 92.)

Progress is being made. In 1987, the Montreal Protocol on Substances That Deplete the Ozone Layer was signed. Since then, 66 countries have ratified it. But only 20 of them are from developing countries, among them India and China, which have over one-third of the world's population. Like other developing countries, which together produce only 3 percent of the CFCs, India and China cannot afford the higher price of CFC substitutes. They want the industrial countries that created the ozone hole in the first place to assist them in devising non-CFC technologies.

Despite widespread agreement among the treaty signatories, however, the U.S. until recently has opposed the creation of any such fund. The fear was that it would set a precedent of providing environmental aid money to developing countries and would lead to more demands to help face such looming problems as global warming. Our failure (currently 1990) to share in the proposed fund risks the future of the ozone layer. Should the U.S. support such a fund?

Drawing by D. Fradon; © 1988 The New Yorker Magazine, Inc.

Issue 31

Hazardous Chemicals, Dioxin, and Molecular Formulas

Currently about 50,000 different chemicals are being produced by U.S. industries. It takes the Environmental Protection Agency three years and $250,000 to study thoroughly the possible long-term toxicological effects of just one of them. At this rate, it will take 150,000 years to study the chemicals we have now—and a new chemical is being introduced into the biosphere every 20 minutes (1988).

Drugs and pesticides are the most fully tested of all chemical products coming onto the market. Yet, for only about 10 percent of those in use today do toxicologists have relatively complete information on the health hazards they create. Food additives and cosmetics are even less well tested (2 percent). No tests at all are performed on the toxicity of nearly 80 percent of such general chemicals as solvents, paints, and plastics.

Meanwhile, the Consumer Product Safety Commission reports that even some chemicals found on the school chemistry laboratory shelf *may* be carcinogenic (cancer-causing) if used in sufficient quantities over a period of time. These include benzene, cadmium nitrate, carbolic acid, carbon tetrachloride, ferric oxide, chloroform, chromium nitrate, formaldehyde, kerosene, phenol, propanol, and tannic acid, among others.

Particularly dangerous is dioxin, the general name for a class of molecules created by reactions between oxygen and cyclic chlorinated hydrocarbons at high temperatures. Dioxin is an unwanted by-product in the manufacture of various herbicides (e.g., 2,4,5-T). It is also one of the most toxic substances known after the botulism toxin. As an accidental contaminant in the oil that was sprayed on roads by a waste hauler a few years ago, dioxin forced the permanent evacuation of hundreds of people from their homes in Times Beach, Missouri.

Analyzed, 1000 grams of dioxin produces, in part:

— 0.4410 g chlorine
— 0.4472 g carbon
— 0.0994 g oxygen
— 0.0124 g hydrogen.

When vaporized, and its vapor density compared with the vapor density of a known gas, the molar mass of dioxin is found to be 322. What is the molecular formula of dioxin?

As long as society lawfully sanctions the manufacture of toxic chemicals, there must be dumps created in which to dispose of them. Yet, few communities acknowledge any responsibility to share this burden. There is no public ethic that imposes it on them. Instead, the NIMBY ("Not in my backyard") attitude determines the response of voters in nearly every candidate community. For this reason, no major new hazardous waste dump has been sited in the U.S. since 1980.

To avoid future tragedies, who should be responsible for the future disposal of such substances: the producer of the substance? the waste hauler? a state agency? a federal agency? How should the cost be handled? Give reasons for your answer.

(See "Oil Pollution," p. 64; "Underground Toxic Wastes," p. 59; and "Waste Disposal," p. 57.)

Think about this . . . Should people who believe in astrology be allowed to teach science in public schools, practice medicine, or hold public office? Would you vote for such a person if you had the choice?

Issue 32

Manganese, Platinum, Chromium, and Foreign Policy

The chemistry of the elements manganese, platinum, and chromium bears directly on our national policy toward South Africa.

Our government's opposition to that government's suppression of civil rights is inevitably complicated by the fact that we do not know how to harden steel without manganese and that South Africa has 71 percent of the world's supply. (The Soviet Union has 21 percent.) South Africa has 81 percent of the world's platinum ore. It has 84 percent of all the world's chromium too, which is an essential component of superalloys used to resist corrosion (in ships) and high temperatures (in aircraft engine turbine blades). Platinum is an active ingredient in catalytic converters, which remove polluting gases from automobile exhaust. (See "Catalysis and City Air Pollution," p. 62.)

Should these facts be allowed to affect our policy toward South Africa? If so, how?

Think about this . . . Should a severely retarded child with an incurable disease be treated when treatment will merely serve to prolong a painful life for a few months? What about 20 years?

Issue 33

Waste Disposal

Each American discards, on the average, 590 kilograms (1300 pounds) of waste material each year. So far in your life, how many tons have you personally been responsible for? Where is it now?

Close to 50 percent the nation's paper, 8 percent of the steel, 75 percent of the glass, and 30 percent of the plastics output is used only to wrap and decorate consumer products. Most of it is thrown away. On your next visit to the supermarket, see if you can find ways of reducing this "throwaway."

Two percent of the nation's trash by weight consists of disposable diapers (which are not biodegradable in a landfill). That's 18 billion diapers per year, nationwide.

The Fresh Kills Landfill on Staten Island, New York City, is an imposing pile of garbage 42 meters (140 feet) high. It receives 20,000 metric tons (22,000 tons) of garbage daily and should reach a projected height of 150 meters (500 feet) by the year 2000, its scheduled closing date. When that time comes, it will be the highest summit on the Atlantic Coast between Maine and the tip of Florida. *There,* indeed, is a symbol of our nation's wastefulness! (See "Law of Conservation of Matter," p. 46; "Hazardous Chemicals, Dioxin, and Molecular Formulas," p. 54; and "Underground Toxic Wastes," p. 58.)

Think about this . . . *Is the dramatic increase in milk production made possible with genetically engineered growth hormones worth the unknown risk to children's health?*

Issue 34

Underground Toxic Wastes

Are you being poisoned by your drinking water?

From as many as 16,000 landfills around the U.S., pesticides, human-made organic chemicals, heavy metals, and other poisons are seeping into the earth and into the water table. Add to that buried but leaky gasoline tanks, septic systems, and runoff of farm fertilizer. It should not be surprising that up to one-fourth of the nation's aquifers (underground natural freshwater reservoirs) will be contaminated beyond use in coming years.

In California's "Silicon Valley," solvents used in making computer chips have seeped from buried storage tanks into the water supplies of several communities. Ethylene dibromide, a carcinogenic fumigant, has contaminated wells in Florida. It may cost up to $1.8 billion to completely clean up leaking pesticides and nerve gas residues at the Rocky Mountain Arsenal near Denver, Colorado.

In 1986, 26 companies were developing genetically engineered crops that resist herbicides so that farmers can use herbicides to clear their fields. But the herbicides are already showing up in the aquifers of ten farm states.

Price's Pit, a 9-hectare (22-acre) dump for toxic wastes located 10 kilometers (6 miles) north of Atlantic City, New Jersey, was the repository for 34 million liters (9 million gallons) of toxic chemical wastes dumped between 1971 and 1973. In nearby Egg Harbor Township, the tap water blackened pots, turned laundry yellow, and at times fizzed like soda pop. The pollutants seep about 18 centimeters (7 inches) a day, or half a mile in 10 years. Already they have reached the Cohansey Aquifer, which supplies Atlantic City with water. Last year, over 33 million people visited Atlantic City.

What should be done? By whom? And who pays?

(See "Hazardous Chemicals, Dioxin, and Molecular Formulas," p. 54 and "Waste Disposal," p. 57.)

Issue 35

Radioactive Waste Disposal

How do you feel about something that has to stay underground and be sealed off for 25,000 years before it is harmless?

The disposal of long-lived radioactive fuel rods and other waste products from the nation's nuclear reactors and medical community poses two serious problems. The first problem is how to protect people from the radiation during the thousands of years it remains dangerous. The solutions now being studied involve forming the wastes into various forms of inert and insoluble glass or artificial rock and burying it deep in geologically stable regions of the earth.

The second problem is to decide just where the burial sites are to be and how to pay for them over their long lifetimes. In your town, perhaps? Why not?

In a democratic society like ours, waste disposal is not only an engineering problem, it is a political problem too, requiring consideration by all segments of society. This difficult problem must be solved before *anything* happens. Meanwhile, the radioactive wastes continue to accumulate.

A thought question or class project: Where in your state is the most practical place for the disposal of the state's radioactive waste? Ideally, the answer must take into account the geology, politics, transportation routes, population density, jobs, water table, and the volume of radioactive waste to be received and stored from the state's hospitals, industry, and nuclear power plants. Is your choice compatible with present state laws?

A related problem: 10,000 years from now, when the wastes will still be dangerous, the English language may have evolved beyond recognition. How do you record safety instructions and precautions *on the tanks* for the people of that distant time? Records on paper elsewhere might be lost.

Should the ocean be considered an option for hazardous radioactive waste disposal?

Issue 36

Acid Rain

Rain may be bad for you!

Factory and power plant smokestacks emit a variety of chemicals, particularly nitrogen oxides and sulfur dioxide, that make the downwind rain acidic. For example (simplified):

$$2SO_2 + O_2 \rightarrow 2SO_3 \text{ and then}$$

$$SO_3 + H_2O \rightarrow H_2SO_4 .$$

The rain, turned into sulfuric acid (H_2SO_4), nitric acid, and hydrochloric acid may be as concentrated as the acetic acid in vinegar,. It falls on lakes, farms, cities, and people. It kills fish, creates lung disorders, damages buildings made of limestone and marble ($H_2SO_4 + CaCO_3 \rightarrow H_2CO_3 + CaSO_4$), rusts autos, and stunts plant growth.

The Parthenon in Athens and the architectural treasures of Venice, as well as numerous stone buildings in the U.S., are being destroyed steadily by acid rain from upwind factories. The factories employ many people. Should the factories be allowed to expand in order to remain economically competitive?

Some corrosion products of acid rain also contaminate drinking water because copper, lead, and zinc pipes all dissolve slowly in acid rain,

$$Zn + H_2SO_4 \rightarrow ZnSO_4 + H_2 .$$

In Latin America, many houses have galvanized sheet iron roofs from which rain is collected for drinking water. In areas of active volcanoes, the rain is made acidic by the volcanic gases. It dissolves the zinc coating on the sheet iron and poisons the drinking water. Zinc poisoning is common in some of these areas.

In the Netherlands, corrosion by acid rain has thinned the walls of bronze church bells. Since the thickness of a bell's wall determines the tone, its pitch is irreversibly altered. Carillon bells normally in tune for 300 or 400 years are completely out of tune in 25 to 50 years.

An epidemic of green-tinted hair among blond and white-haired residents of Columbia, Maryland, was traced to the drinking water. Their well was so highly acidic that it was dissolving copper from the pipes, which they then ingested. The copper caused the change in hair color.

Lakes and forests are severely affected by acid rain too. Hundreds of lakes in Ontario and the Adirondack Mountains are devoid of all fish, and thousands are damaged. About 9000 are threatened east of the Mississippi. Acid rain has destroyed all aquatic life in 10,000 lakes in Sweden. In Europe, 22 percent of the forests are showing signs of damage, and some are totally destroyed.

In 1988 President Reagan agreed to freeze U.S. nitrogen oxide emissions at the 1987 levels until 1996. An international acid rain treaty is now (1989) being negotiated. Any laws that may be passed requiring factories to install expensive scrubbers in their smokestacks, however, will take 10 or 15 years to have an appreciable effect on the acidity of rain in the Northeast and may cost many jobs.

Who should pay for any cleanup required by legislation? What consequences (political, economic, environmental) do you think will follow if (a) corrective action is taken promptly and (b) action is not undertaken promptly? Should the people living around a Midwestern power plant have to pay to clean up the rain in eastern Canada?

Is it right that pollution-control laws should force a company to go bankrupt or cause workers to lose wages or jobs? Should the U.S. replace its coal-burning plants with nuclear power plants? Should you be restricted from driving your air-polluting car into a nearby city?

If you are a senator or representative for a state in which much of the pollution originates and immense numbers of jobs and dollars are at stake, what are you going to do?

Think about this . . . *If a few aging aircraft suffer explosive decompressions, should all old airliners be grounded?*

Issue 37

Catalysis and City Air Pollution

In traffic-congested large cities, even nonsmokers suffer lung damage from air pollution. Occasionally. the damage may be equivalent to a rural dweller who smokes a pack of cigarettes a day. Joggers are especially at risk.

The pollutants come mainly from automobile exhausts. For example:

$$C_8H_{18} \rightarrow 2C_2H_4 + C_3H_6 + CH_4$$

(octane yields ethylene + propylene + methane), and also:

$$C_8H_{18} + 11\frac{1}{2}\,O_2 \rightarrow 6CO_2 + 2CO + 9H_2O$$

When these and other hydrocarbons and nitrogen oxides mix in the presence of ultraviolet radiation in sunlight, they form ozone and other noxious compounds collectively called *smog*. Smog may contain 100 different compounds, many of them irritant, toxic, or carcinogenic.

The principal method of reducing the emission of these noxious products is catalytic converters in the exhaust systems of cars. A converter contains 1 to 3 grams of a mixture of platinum and palladium metals embedded in a base of aluminum oxide. This mixture catalyzes the conversion of the polluting carbon monoxide and hydrocarbons to carbon dioxide and water. But platinum (up to $600 an ounce, mainly from South Africa) and palladium are very expensive. The unleaded gasoline that must be used in cars with converters is more expensive than ordinary gasoline. Also, the use of a catalytic converter reduces a car's gas mileage. Some people disconnect the converter in their cars. Should they do so? Would you do so? Why?

The nearly 400 million automobiles in the world today emit about 496 million metric tons (547 million tons) of carbon into the atmosphere annually. At that rate, if nothing is done, these emissions will nearly double by 2010, increasing global warming and city smog. The increased American demand for gas-guzzling cars in 1988 was clearly of no help. (See "Manganese, Platinum, Chromium, and Foreign Policy," p. 56.)

While gasoline engines are the major source of smog in cities, solar ultraviolet light catalyzes other sources of various gases to ozone and smog. Gas stations, hamburger restaurants (the smell of sizzling meat!), paints and varnishes, bakeries, dry cleaners, breweries, lawn mowers, and backyard barbecues are all sources of smog. Their uses are deeply embedded in the habits of Americans. Officials of the Environmental Protection Agency say as many as 75 or 80 areas of the country fail to meet federal Clean Air Act standards, and their compliance will require a slash in offending emissions by at least 50 percent.

If your lifestyle or your livelihood depends on any of these industries or activities, how would you react to such severe regulation? Would you fight it? How? Evade or ignore it? Move away? What? How much right of protest have you? Is this an issue on which to exercise it?

Ozone in the *upper* atmosphere is created differently from ground-level smog, although human-made chemicals and lifestyle play an essential part in its destruction. (See "Ozone Catalysis: Destruction of Our Atmosphere," p. 52.)

It has been suggested that smog may be protecting people from the increased solar ultraviolet radiation that results from the destruction of upper-atmosphere ozone. If so, one source of pollution has offset the bad effects of another! Should this be cause for rejoicing or complacency?

"They have very strict anti-pollution laws in this state."

Drawing by Richter; © 1989 The New Yorker Magazine, Inc.

Issue 38

Oil Pollution

Should a nation (Iraq), heedless of long-term damage to wildlife in the enclosed waters of the Persian Gulf, be charged in some way for deliberately releasing enormous volumes of oil as a defense against the 1991 UN-sponsored invasion to capture Kuwait?

Should an oil company, negligent of legal safety precautions, be charged for criminal damage if an oil spill pollutes a public park or wildlife refuge or harbor (e.g., the Exxon *Valdez* spill in Alaska in 1989)?

How about the individual company employees whose negligence was responsible (e.g., the captain of the Exxon *Valdez*)?

Should you be charged with a criminal offense if you dispose of polluting material (oil, detergents, old paint, etc.) into a park or a waterway?

Is your answer the same for simply pouring the polluting material down the drain?

Is there a moral distinction between you and the oil company?

(See "Hazardous Chemicals, Dioxin, and Molecular Formulas," p. 54.)

Think about this . . . *Trail bikes and dune buggies on public lands destroy fragile vegetation and the peace and quiet. Should these vehicles be outlawed? Loud motorcycles on public highways too?*

Issue 39

Petroleum Depletion

Within the next 10 or 15 years, the decreasing supply of petroleum will start to fall short of the world's demand for it. As petroleum reserves decline, alternatives to gasoline become more attractive and interesting. One is methyl alcohol (methanol), made synthetically from hydrogen and any carbonaceous material that can yield carbon monoxide.

$$CO + 2H_2 \rightarrow CH_3OH \quad \text{(with catalyst)}$$

Mixed with gasoline, it's called "gasohol." Any present gasoline engine, including diesel, will run on it. It's already cheaper than gasoline. Problems: greater engine wear and such exhaust pollutants as formaldehyde, CH_2O.

Other processes are the extraction of petroleum from a form of shale, and the Lurgi process (steam and oxygen over hot coal makes octane, among other things), but they are both expensive and polluting.

It is not widely appreciated that petroleum is needed to make plastics, synthetic rubber, and artificial fertilizer. For these important processes there are no satisfactory substitutes. A good case may be made for trying to preserve the world's remaining petroleum for these important uses by finding alternatives to petroleum (such as those mentioned above) or developing alternative energy sources or conservation.

The complex interactions of geology, politics, and economics will dictate a drop in petroleum production in the next two decades. Can you write a scenario for the sequence of events as this begins to happen?

Think about this . . . *Smoking kills and damages the health of more people than drugs do. So, shouldn't tobacco advertising be made illegal? Or drugs legal?*

Issue 40

Ocean-Bottom Minerals

Vast wealth for the taking?

In the ocean, far out beyond national boundaries, large areas of the seafloor are covered with marble- to baseball-sized lumps rich in the valuable metals copper, manganese, nickel, and cobalt. No nation owns the oceans; they are international. But only the developed industrial nations have the money and

technology to harvest this wealth, which belongs to *all* nations. *Should* it be harvested? How are the poorer nations' rights to be protected?

The U.S. has refused to ratify the 1982 U.N. Convention of the Law of the Sea, which sought to regulate mining and other commercial development. The Administration argued (in 1988) that the treaty, which assigned ownership of the resources on the deep-sea bottom to all nations, interferes with private exploitation. The U.S. Senate has yet to ratify a similar treaty of 1979 concerning resources on the moon and other celestial bodies. Thus, those resources are available for you or anyone else to take who can afford to exploit them.

But why should minerals on the *bottom* of the ocean beyond the 32-kilometer (20-mile) limit be any different than the fish *in* the same ocean merely closer to the surface? Persons with resources for big fishing operations take more fish than the hook-and-line fishers, and so it has been for ages.

Consider in this same discussion the use of 64-kilometer-long (40-mile-long) drift nets that snare not only all the fish to a depth of 30 meters (100 feet) or more but also ensnare dolphins, sea turtles, and whales, killing them and sweeping an immense volume of the sea clean of all organisms larger than the mesh size of the nets.

By what means may destructive exploitation of this common heritage of humankind be brought under control? Is it possible to do so and still have it freely available? (See "Tragedy of the Commons," p. 110, for which all the above resources provide examples.)

(See "The Geopolitics of Genes," p. 18.)

Think about this . . . In our culture, why is nature ("Mother Nature") feminine—and something to be conquered?

Issue 41

Precipitation: Cloud Seeding

Sprinkled into some kinds of clouds, dry ice (solid carbon dioxide) or crystals of silver iodide create immense numbers of ice crystals, and often rain. This may be good for parched farmlands and depleted city reservoirs, but such cloud seeding may be a disaster for recreation areas and may even lead to damage by flash floods.

The Down Side to Cloud Seeding

On Buffalo, New York, downwind from Lake Erie, over 250 centimeters (100 inches) of snow are often dumped each winter. Cloud seeding over the lake should result in the snow being deposited *in* the lake *before* reaching the city—or at least spread over a larger area. Would you vote for an experiment in cloud seeding if you were:

— the mayor of Buffalo?
— the owner of a big ski resort a few kilometers north of Buffalo?
— manager of the Buffalo airport?
— superintendent of Buffalo city schools?
— Erie County roads commissioner?

Who should decide where and when to seed? Will it work on hurricanes? If serious damage is caused by a "seeded" storm, should the people doing the seeding be held responsible?

Think about this . . . *If a major oil deposit were found beneath one of our national parks, should it be exploited by drilling?*

Physics

Piecemeal social engineering resembles physical engineering in regarding the ends as beyond the province of technology.

— SIR KARL POPPER
The Poverty of Historicism

Issue 42

Newton's Laws and the Existence of God

In the thirteenth century, St. Thomas Aquinas proposed five proofs for the existence of God in his *Summa Theologiae*. His first proof begins with the observation that no object can be made to move except by the action of another. But there cannot be an infinite series of successive movers, he says, and, therefore, there must be a first or prime mover that is not moved by anything else—this is God.

Is St. Thomas's argument consistent with Newton's (seventeenth century) laws of motion? If not, why not?

Think about this . . . Does science tend to break down people's ideas of right and wrong?

Would you vote to outlaw the sale of irradiated food in food markets—or be willing to eat it? Compare with iodized salt and fluoridated drinking water.

Issue 43

Newton's Second Law ($F = ma$) in Athletics

Without protective padding, a prize fighter's wallop or a collision on a football field can exert a force of 1000 pounds for a very brief time. Such a force, F, can deform bones. The acceleration, a, it produces in limbs, trunk, or skull can twist, stretch, or rupture nerves and blood vessels. $F = ma$ may be written as

$$F = mv/t \text{ or } Ft = mv$$

where m and v describe the fighter's fist or the football player's body as the collision takes place. Newton's second law!

Protective clothing spreads the action of the force over a longer period of time, t. In the product Ft, if t is increased by padding, the average force, F, must decrease (since the product Ft equals the unchanged and therefore constant mv), thereby reducing the resulting bodily damage.

Cyclists, motorcyclists, construction workers, football players, and hockey players are protected by helmets because $F = ma$.

Professional boxers do not wear helmets, and their padded gloves weigh only 8 ounces. Should laws require them to wear gloves heavier than 8 ounces?

Are jogging shoes designed with $Ft = mv$ in mind? In what ways?

Why do elastic poles permit higher pole vaults than the older rigid poles?

Why do automobile seat belts reduce injuries? A passenger's mv is the same in a collision with or without a seat belt.

A majority of states now have mandatory child-restraint laws for automobiles.

Issue 44

Elementary Mechanics
and Women's Liberation

Women's liberation has its roots in applications of elementary mechanics and electricity—notably in the invention of the sewing machine, the typewriter, and the telephone.

If you had been a schoolgirl in America a hundred and fifty years ago, clothing stores would not be a part of your world. None existed. Only wealthy people could afford professional tailors to make their family's clothes. More commonly, a farmer's or a worker's wife spent 4 to 6 hours *daily* making and mending the family's clothes. Elias Howe's invention of the sewing machine in 1846 eventually reduced that time to more-nearly half an hour a day and made it possible to create clothes so inexpensively that average-income people could afford to buy clothes in stores. Try to image the effect this one invention would have had on your daily life!

But even with a sewing machine to ease your labors, a girl's life during the Civil War would still have been limited to the home. In Dickens' last novel (around 1870), "respectable" women did not go outdoors without an escort, and only male clerks worked in offices. (Compare that with the lives of women in Moslem countries today.)

Only 25 years after Howe's sewing machine, the invention of the typewriter (about 1867) and the telephone (Alexander Bell, 1876) had changed all that, and by 1900 women were travelling alone outdoors and working and leading lives of their own outside the family circle. These changes created and led to a need for higher education, once considered a luxury. And that, together with a demand for the vote, equality before the law, and for equal access to careers, followed inevitably.

If such simple applications of science can have such profound effects on the lives of so many people, what effect is the invention of the computer having on your life and your job future? Talk to people whose jobs 25 years ago had no computers.

Drawing by Handelsman; © 1989 The New Yorker Magazine, Inc.

Issue 45

Accelerated Motion Equations and Highway Safety

During the 1973 energy crisis, federal law mandated a reduction of the speed limit from 65 mi/hr to 55 mi/hr on the nation's highways to save gasoline. The accelerated motion equation:

$$v^2 = 2as$$

provides one of the reasons why the annual death toll went down by 27 percent.

The equation says that if you go twice as fast (v doubles), it takes your car *four* times as much distance (s) in which to stop. Conversely, if you slow down from 65 mi/hr to 55 mi/hr (0.85 times as fast), you can stop in 0.72 (or less than three-quarters) as great a distance. The difference saves lives.

In April 1987, the federal government gave each state the power to raise the speed limit from 55 mi/hr to 65 mi/hr on its rural interstate highways. The National Research Council reports that if this increase takes place nationwide, it will cause 2000 to 3000 additional traffic deaths annually.

Imagine you are a member of your state legislature, voting on whether to raise the speed limit. Knowing the cost in lives, which way would you vote?

Note that approximately 45,000 Americans are killed every year in automobile-related accidents. That is roughly equivalent to one fully loaded passenger airplane crashing each day.

Do you approve of motorists using radar detectors ("fuzz busters") to frustrate highway police speed control?

Should the use of automobile seat belts be mandatory?

While lower speeds save lives, seat belts save lives too. When a car going 80 km/hr (50 mi/hr) hits a wall, the car stops in about 0.16 sec. But the humans inside it (if they are not wearing seat belts) keep going at 80 km/hr and collide with the inside of the car 0.03 sec later. This second collision is equivalent to diving from a third-floor window head first with only your arms available to stop your fall at the bottom. Is it true that "speed kills"?

Experience with airplane crashes shows that lives could be saved if the seats, instead of facing forward, face the rear of the plane. This arrangement would make maximum use of seat padding and head support. The view out the window at 3050 meters (10,000 feet) would hardly be changed. In what ways would this change be good or bad physics? Good or bad for business?

The kinetic energy equation, K.E. = $mv^2/2$, says that at 10 times the speed a collision can do *100* times the damage. Compare the damage from hitting a big tree when jogging at 10 km/hr (about 6 mi/hr) with the damage when driving at 100 km/hr (about 60 mi/hr).

(See "Newton's Second Law ($F = ma$) in Athletics," p. 73.)

Think about this . . . Wealthy people can afford to pay for heart transplants, so Medicare or Medicaid funds should be available to pay for heart transplants for only those who are poor. Do you agree?

Issue 46

Science and Athletic Records

The laws of science have always had a lot to do with athletics. The understanding of those laws continues to result in improved performances and steadily broken world records. Consider the idea of force. The study of how forces act makes it clear that:

— the dimples on a golf ball make it travel more than 4 times farther than if it were smooth.
— a running track with a springy surface about twice as stiff as a runner's legs is the ideal for improving speed.
— the best path for a swimmer's hand through the water is an S-curve, not a straight line.
— the best way to keep an Indy-500 race car firmly on the track is with the help of stubby boxes shaped like airplane wings turned upside down.
— ski jumpers travel farther if they stretch their bodies low over their skis to gain lift like an airplane wing.
— floor oil on the first 3 or 5 meters (10 or 15 feet) of a bowling alley reduces the force of friction and so preserves the ball's spin until it is close to the pins, thus improving the bowler's control.
— a properly shaped and thrown boomerang can stay in the air as long as 33 seconds (world record).
— tennis racquets made of graphite, which is 10 times stiffer than wood, result in more powerful shots; Fiberglas poles resulted in an immediate increase of 23 centimeters (9 inches) in the world's pole-vaulting record.
— skin-tight clothing that reduces a runner's air resistance can mean a difference of as much as 10 centimeters (4 inches) in the 100-meter dash and 27 meters (30 yards) in a marathon. It improves a cyclist's performance too.

Compare the structure of the human body with that of (a) an antelope for speed, (b) an elephant for carrying weight, (c) a bird for flying, and (d) a monkey for climbing. Each of these four-limbed animals is superior to us in a very specific way. Being so badly outclassed by them, how is it that humans are the most successful large animal on earth?

Further use of the various laws of physics, chemistry, and biology continues to push human performance to still greater limits.

__Think about this__ . . . Where is the best area in your state on which to "go back to the land," raising your own food, living simply, and becoming as nearly independent as possible from the tools, amusements, and energy sources of twentieth-century America? If the pioneers did it, couldn't you do it today?

"Birdie, do you think that if I had a suit and shoes like hers, my chances of survival would improve?"

Issue 47

Heat, Light, and Sound

Is your home entirely your own?

Solar home heating and solar water heating have become economically practical in many parts of the country. Imagine that your house or apartment has been equipped with one or both of these. A tall apartment building is put up next door, blocking half the sunlight. Should you have any legal recourse? (For half the sunlight, substitute one-quarter—or all.)

The apartment building has tennis courts alongside. The courts are lit at night, lighting your bedroom brilliantly and making it hard for you to sleep. Should you have any legal recourse? (For tennis courts, substitute a shopping center or a mobile home park.) (See "Scattering of Light: Light Pollution," p. 89.)

The nearby airport adds a runway that leads planes directly over your house at take-off, making conversation and sleep difficult. The maximum loudness is 100 decibels, 10 times a day. (See "Loud Sounds," p. 87.) Should you have any legal recourse?

Finally, close to the other side of your house, oil (or other valuable mineral) is discovered on your neighbor's land. The deposit presumably extends underneath your home at a depth of 3 meters (30 meters, a kilometer), but you do not want it mined. How deep into the earth does your ownership extend?

If you argue that the homeowner has a right to stop each of the four threats to his or her comfort, consider each situation from the opposite point of view. Should your plans to build a big apartment house be blocked by the existence of one neighbor's modest solar panel? Or your plans to play tennis at night be blocked by someone else's desire to have a bedroom on the same side of the house? Should the airport of an entire town be thwarted by one family's dislike of noise?

In resolving any of these questions, the measurement of heat, light, or sound is important.

Winston Churchill once said: We shape our buildings and then they shape us.

(See "High-Voltage Power Lines," p. 90.)

Think about this . . . If we establish radio communication with an extraterrestrial civilization, what do you think we should say to it?

Should developing nations that are overpopulated use compulsory sterilization to control their populations?

Issue 48

Air Conditioning

After World War II, engineers developed techniques for cooling large volumes of air below its dew point to condense and remove its water vapor inexpensively. Mechanical air conditioning was born, and it revolutionized life in the South.

In the 1960s, for the first time since the Civil War, the South experienced a net in-migration and the prosperity associated with the Sun Belt began. Along with prosperity came isolation and a loss of regional flavor. It can be argued that, along with other manufacturers of air conditioners, General Electric Company has had a more unsettling effect on the South than General Sherman did during the Civil War.

In what ways is an electric fan better than an air conditioner? How do people elsewhere in hot, humid climates manage to stay comfortably cool?

Would the widespread use of air conditioners be a sensible response to an impending greenhouse effect? Why? (Consider how an air conditioner works. It uses CFCs. As it cools indoor air, it heats outdoor air.)

Think about this . . . Should scientists be forbidden to do research on some subjects, such as nuclear explosions, powerful insect poisons, methods of altering human genetic material, and the means of warfare?

Issue 49

Cryogenics

Many lower animals can be quick-frozen into a solid block and maintained alive at very low temperatures until they are restored to good health by thawing (e.g., goldfish dropped into liquid air, about −195°C). Suppose it becomes possible to freeze and thaw humans. If you have an incurable disease, might you have yourself frozen until a cure has been found? Should public funding be made available for this procedure, as it is for other medical procedures?

"They say I have my grandmother's eyes, my grandfather's nose, and my great-great-Auntie's chin!"

Already, such human tissues as blood, corneas, bone marrow, sperm, and even embryos can be frozen and then thawed later. So far, however, transplant surgeons have been unable to make use of such frozen organs as kidneys, the pancreas, or the heart, which therefore makes it impractical to freeze and then thaw entire people. (At least five people, though, were known to have been frozen in liquid nitrogen in 1988, and possibly as many as twelve or more.) They face an uncertain future, for when they are thoroughly frozen their heartbeat and their brainwaves stop. By definition they are legally dead for as long as they remain frozen. That could be for years or centuries!

If you have been frozen and it takes 100 years for a treatment for your disease to be found, should your great-great-great grandchildren have to be responsible for you when you are thawed out? If you left a will disposing of your worldly possessions, could you reclaim them? Would this practice be ethical to perform on (a) astronauts making years-long space flights, (b) depressed people as an alternative to suicide, (c) the curious who simply want to see what the future will be like without growing old while waiting?

If people who had been frozen in 1800 were revived successfully today, what problems would they face? What would their age be when they are thawed out? For that matter, how old is the thawed-out goldfish? Consult an English teacher on the fate of Rip Van Winkel.

Think about this . . . *Would two-way television threaten your Constitutional right to privacy?*

Issue 50

Convection

Hundreds of years ago in primitive societies in cold climates, people kept warm by huddling together around an open fire in the center of their living quarters. Above the fire, a hole in the roof let out some of the smoke and a good deal of the heat.

In the ninth century in Europe, it was found that a properly designed fireplace and chimney created a convection current that carried away all the smoke and, by radiation from the fireplace walls, increased the heat to the room. Small fires could be built *anywhere* in a multi-story building, which encouraged the subdivision of living quarters. Privacy and social divisions were enhanced.

It has been said that the invention of the chimney may have affected the art of love in medieval Europe more than the troubadours did. It certainly isolated a lord and his lady in their "withdrawing room" from the rest of the household, increasing class divisions.

In William Langland's *The Vision of Piers Plowman* (fourteenth century):

Woe is in the hall each day of the week.
There the lord and lady like not to sit.
Now every rich man eats by himself
In a private chamber to be rid of poor men,
Or in a chamber with a chimney
And leaves the great hall.

Has modern central heating reversed the fragmentation of a family household? Is television a modern parallel?

(From an untitled review article, Lynn White, Jr., *Isis* (75) 1:276 [March, 1984], p. 172 et seq.)

The family refrigerator is another application of technology that has had profound effects on family life. Its invention has made daily trips to the corner market (as our grandparents did) unnecessary; it also has kept us from having to make contact with our neighbors. Taking away a common ground on which people used to meet daily, the refrigerator has helped isolate us even more from one another—as television does.

Think about this . . . *"For profit" hospitals are basically businesses. Should they be required to accept indigent and uninsured patients?*

Should citizens, although not specially trained in science, have the authority to vote to halt scientific research that is believed to present possible hazard to human life?

Issue 51

Loud Sounds

With the exception of thunderclaps, most loud sounds are human-made. Many of them damage your hearing permanently in a rather short time.

For example, a jet engine close up, rock concerts, and some stereo headsets at 135 decibels damage your hearing instantly. A big motorcycle with no muffler, or an average chain saw at 110 decibels, takes 10 or 15 minutes to cause damage. A jackhammer or a big power mower at 100 decibels will cause damage in 1 to 2 hours. At 90 decibels a convertible ride along a busy freeway takes 4 hours to do damage. A screaming child can do hearing damage in 8 hours. Alarm clocks and vacuum cleaners around 75 decibels have no time limit. Not surprisingly, by the age of 40 most Americans have lost some hearing in the top two octaves of their hearing range.

In January of 1988, the Denver Broncos played the Cleveland Browns in a play-off game in Denver. The *Rocky Mountain News* reported that a sound level meter read a peak of 112 decibels when Denver recovered a Cleveland fumble. For a news note on a $27,000 "boom-car" system that topped 150 decibels, see "Shake, Rattle, and Roll" in *Time,* March 6, 1989, p. 52.

When Kent Hrbek hit his grand-slam home run in the sixth game of the 1987 World Series, the sound level inside the Metrodome in Minneapolis reached 120 decibels! ("120 Decibels!" letter from Thomas D. Rossing, *The Physics Teacher,* Dec. 1989, p. 644.)

Even when it is not deafening, noise causes stress and is linked to headaches, ulcers, and high blood pressure. An unborn child responds to sudden loud noises with increased heartbeat or by kicking.

In a New York City classroom adjacent to elevated train tracks, the reading ability of a class of sixth graders was found to be a year behind that of sixth graders on the quiet side of the school building. Reading scores became the same on both sides after remedial steps were taken.

What is noise? How is it different from music? Can music ever be considered noise?

Does your community have noise regulations? If so, how is the noise level measured? What are the penalties for excessive noise? Do you consider them to be reasonable?

Think about this . . . Should our nation's space program include plans to establish an inhabited base on the moon or Mars? Why?

Would you be willing to take a job working on a product (e.g., robots) that you knew would put people out of work? What about working in a gun factory?

Issue 52

Scattering of Light: Light Pollution

Have you ever seen the Milky Way?

Like the Milky Way, Halley's comet, when it returned in 1985 and 1986, was invisible or badly dimmed for many people because of light interference from our brightly lit cities and suburbs.

The scattering of city lights by overlying smog particles gives the night sky a milky color against which even the brightest stars are often invisible. Because of such "light pollution" many high school students now grow up never having seen the Milky Way spanning the night sky.

Already the usefulness of the nation's largest optical telescope, the 200-inch instrument on Mt. Palomar in California, is being threatened seriously by the increasing glare of the lights of San Diego. Some telescopes on Kitt Peak, in Arizona, are being limited by the lights of Tucson. The famous 100-inch telescope on Mt. Wilson, in California, is now useless and is being retired because of the lights of Pasadena.

Even amateur astronomers looking at the night sky through small telescopes are having more and more difficulty finding places and times where the sky is dark enough to see planets, comets, stars, and nebulae. The moon, at least, is almost always easy to see.

If nothing is done, what will be some of the long-range consequences?

Note: Mt. Palomar astronomers named a newly discovered asteroid "San Diego" as a "thank you" to the city for its cooperation in shielding its street lights.

(See "Outer Space: Infinite Dump," p. 106.)

Issue 53

High-Voltage Power Lines

High-voltage power transmission lines serve essential human needs and make modern cities possible. They create human problems too. *Powerline* (by B. M. Casper and D. Wellstone, Univ. of Mass. Press, 1981) describes in vivid detail the reasonableness of the plans for an 800,000-volt power plant (d.c., incidentally, not a.c.) and transmission lines to carry energy from a coal field in North Dakota, across 690 kilometers (430 miles) of Minnesota farmland, to Minneapolis and St. Paul. The farmers whose fields lay in the path of the power line had reasonable objections. The strenuous—and violent—conflict that resulted is a vivid example of citizen protest against what was perceived as the unreasonable encroachment of technology on private property and voters' rights.

How would you react to a high-voltage power line being built 100 meters (109 yards) or less from your home? On what grounds might you object? To whom? (For a further strong view on power transmission lines, see the quotation by Thomas A. Edison in *Quotations* at the back of this book.)

A related issue is the perceived hazard of the electromagnetic fields surrounding power lines and electrical appliances in the home. What are their possible adverse effects on the health of people living near them? Of the two, home wiring circuits, lighting fixtures, and such appliances as TVs, computer terminals, and electric blankets may play a greater role than transmission lines in posing a public health problem.

Laboratory experiments on living cells and animals have shown a statistical association—but no cause-and-effect relationship—between cancer and exposure to electromagnetic fields from wires that carry electricity through neighborhoods and throughout homes. With no risk standards for avoiding exposure or proof that any are necessary, how should you proceed? Do you avoid the television set? Or an electric blanket (which is particularly close to you for an extended time)? Or other electrical appliances?

Issue 54

Current Electricity and the Telephone

In July 1986 a disc jockey in a lighthearted mood announced that the next afternoon the telephone company was preparing to blow the dust out of the telephone lines. He recommended that his listeners put plastic bags over their telephones to protect themselves against the dust. There was a "run" on plastic bags in the local markets as many people hurried to comply.

What does this incident say about people's schooling? Their gullibility? Could this have happened to listeners who had studied current electricity in a physics course? Could it happen in your community?

***Think about this . . .** It is said that by the next century it may be too late to save the environment. In the meantime, what are your priorities for action (**a**) in the world, (**b**) in the U.S., and (**c**) in your community? Are your proposals politically practical?*

Every day there are 235,000 more people in the world than there were the day before. In what ways is this affecting you personally?

Issue 55

Absorption of Radiation: Greenhouse Effect

Just by breathing, you add to the greenhouse effect!

As people burn wood, coal, oil, and gasoline (C + O_2 → CO_2 + heat), carbon dioxide is released into the atmosphere. Like glass panes of a greenhouse, carbon dioxide molecules are transparent to sunlight, allowing it through to warm the earth's atmosphere. But when the earth's surface is warmed, it gives off its heat as infrared radiation to which the carbon dioxide molecules are not transparent. Blocked from radiating its accumulating warmth into space, the earth's surface heats up. As a result of this "greenhouse effect," the surface averages a comfortable 59°F. In the absence of all carbon dioxide, it would be below 40°F, perhaps even below freezing (32°F).

The planet Venus is an excellent example of the greenhouse effect. Venus's 455°C average surface temperature is largely due to its carbon dioxide-laden atmosphere.

Methane (CH_4) emitted by decaying organic material, garbage, and bacteria is 20 times as effective as carbon dioxide, molecule for molecule, and chlorofluorocarbons (CFCs) are 10,000 times as effective in blocking the passage of infrared radiation. Ozone and nitrogen oxides from automobiles are also important contributors to the effect. Another is the accelerating destruction of trees (cutting down forests) and other green ground cover (overgrazing), since their photosynthesis would have *absorbed* carbon dioxide from the atmosphere and thus reduced the greenhouse effect.

$$(6CO_2 + 6H_2O + light \rightarrow C_6H_{12}O_6 + 6O_2).$$

Even the best scientists disagree on whether the effect has already begun. (They *do* agree that the warming is inevitable in the near future.) Their disagreement arises largely because of the complexity of the feedbacks involved (see "Feedback: Positive and Negative," p. 114), which are not yet well understood. It has become clear that we are altering our environment faster that we can predict the consequences. This situation is bound to lead to surprises, mostly catastrophic. Unfortunately, planners may use this lack of unanimity as an excuse to postpone the very unpleasant political and economic actions that soon must be required of them. (See "Long-Term Effects of Technology," p. 116.)

If the warming trend is imminent, Canada and the Soviet Union may one day own most of the world's fertile land. Can you imagine the effect *that* will have on the world's economy and on international politics?

Another consequence of the greenhouse effect would be the melting of the Antarctic ice cap. That would raise ocean levels worldwide as much as 1½ meters (5 feet). The Environmental Protection Agency predicts a 30-cm (1-foot) rise in just the next 30 to 40 years. In parts of Florida, a 30-cm (1-foot) rise may mean 300 meters (1000 feet) of land lost to the sea. Louisiana already is losing 130 square kilometers (50 square miles) of wetlands per year. Wetlands and saltmarshes are the nurseries of the sea; with their loss worldwide, the entire ocean food chain may be broken.

At its extreme, this water rise also will flood seaport cities (e.g., Hong Kong, New York, and Amsterdam) and make them uninhabitable.

As marginal lands are rendered totally uninhabitable because of diminishing water supplies, political conflicts will intensify (e.g., Egypt vs. Sudan over the shrinking Nile River, and between our Western states along the Colorado River). As poverty spurs further depletion of forests and topsoil, hastening the effect, hordes of refugees will be forced to flee Dust Bowl conditions—to where?

And yet, China plans to double its coal production in the next 15 years to spur development!

Another concern is that as temperature and humidity climb, such parasitic and infectious tropical diseases as yellow fever, dengue fever, and Chagas' disease may migrate northward.

What can be done? Consider the pros and cons of each of the following possibilities:

1. Impose special taxes on carbon dioxide emission to encourage energy conservation.
2. Increase funding for research on alternative energy sources, including solar power and safer nuclear reactors.

3. Provide developing countries with financial aid to build high-efficiency power plants.
4. Launch a worldwide tree-planting program.
5. Develop techniques for recovering part of the methane given off by landfills.
6. Require all new cars worldwide to achieve an average of at least 65 kilometers per 3.8 liters (40 miles per gallon) of gasoline and to have catalytic converters for reducing tail pipe emissions.
7. End all industrial production of chlorofluorocarbons (CFCs).

(See "Ozone Catalysis: Destruction of Our Atmosphere," p. 52.)

The "Green" house effect explained:

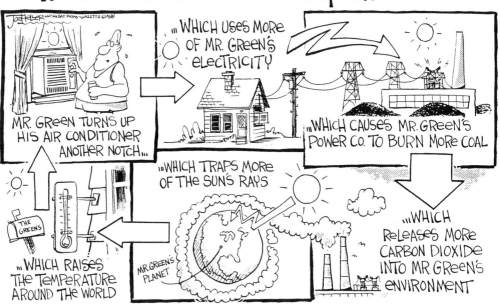

Reprinted with permission of Joe Heller, Green Bay Press-Gazette.

Issue 56

Television: The Third Parent

"**B**oob tube" or "window on the world"?

Television profoundly affects many people's daily habits and their attitudes toward the world around them. From kindergarten through twelfth grade, the average American student spends a total of 11,000 hours in a classroom and 16,000 hours in front of a television. If you fit the average, you may add 10,000 hours more of television-watching for each decade after the age of 20. By the age of 40, the average American has seen more than one million television commercials. In what ways might the rhythm and texture of his or her mind differ from his or her grandfather's?

Think what you can do with that kind of time! Five thousand hours is what a typical college undergraduate spends working on a bachelor's degree. In 10,000 hours, you could learn enough to become a doctor, a lawyer, or an astronomer. You could have learned to speak and read several languages fluently. You could have walked around the world and written a book about it.

Do you think you will want your children to watch this much TV? Why? Do you think *you* should? Why? For many people, television is:

— the most relied-on source of news.
— a creator of political events.
— a primary source of entertainment.
— a way to structure time.
— a national marketplace.
— a major force in bringing up children.

A connection exists between the viewing of violence on TV and aggressive and violent behavior in some young people. George Gerbner, dean of the Annenberg School of Communications at the University of Pennsylvania, reported an average of five or six violent acts per hour on prime time television over the past 17 years of his study (1985). When you have children of your own, are you going to control the programs they watch? If not, why not?

Computers may be as attractive as television for some students; in extreme cases, there are "withdrawal symptoms" a bit like those associated with drug addiction. It seems ironic that the more computers are used in business, home, and school for communication purposes, the less some people communicate.

In your experience, is TV a sedative or a stimulant? What has been the effect of the development of national television networks on:

— regional customs and accents?
— family reading habits?
— awareness of national and international events?
— merchandising of commercial products?
— political campaigns?
— public safety?

"Please, Laura. I know how acid rain turns out, but I don't know who's going to win the ballgame."

Drawing by Handelsman; © 1985 The New Yorker Magazine, Inc.

Issue 57

Nuclear Power: Risk Assessment

Some of the pros and cons for using nuclear reactors as a source of power follow.

Pros:

1. In a few hundred years, known reserves of coal will be depleted. In energy, 1 kilogram of uranium is equivalent to 2,000,000 kilograms of coal. The pressures in favor of nuclear power will become enormous.

WHILE PREPARING A TREAT FOR THE BRIDGE CLUB, MRS. EMILY TROODLE DISCOVERS...

Danzinger in *The Christian Science Monitor* © 1989 TCSPS.

2. Breeder reactors *never* run out of fuel.
3. Nuclear reactors produce no significant emission of greenhouse gases, while coal and oil produce many pollutants dangerous to health and cause acid rain.
4. The problems of spent fuel and waste disposal are being solved.
5. Standardized safer and cheaper reactors are being developed.

Cons:
1. Radioactive poisoning and radiation are deadly.
2. There is no safe and effective way to get rid of radioactive wastes.
3. Stolen uranium and plutonium may be used for weapons.
4. Many nuclear plants are near big cities, where an accident will have particularly severe consequences.
5. Human error was the cause of the Three Mile Island and Chernobyl accidents. Can we ever prevent people from making errors?
6. The impending worldwide uranium shortage will provide strong pressures to reprocess spent nuclear fuel into plutonium, easily diverted into nuclear weapons.
7. Despite the efforts of the International Atomic Energy Agency, Israel and India already have evaded safeguards and joined the nuclear weapons "club." Several other nations are probably close to doing so. Such diversion and proliferation may be impossible to prevent.

Issue 58

Energy: U.S. Consumption

In the course of a year, more energy passes through the windows of buildings in the U.S. than flows through the Alaska pipeline.

The U.S. has only 4.7 percent of the world's population, but it uses annually 31 percent of the world's energy. This rate is equivalent to 59.3 barrels of oil per year per American. Does this rate of consumption carry with it any obligation (a) to the rest of the world's population or (b) to conservation of the sources of energy? What should be done, if anything? What should *you* do, if anything?

Amory B. Lovins, energy conservation advocate and author of *The Energy Controversy: Soft Path Questions and Answers,* (Brick House Pub. Co., 1978) has observed that civilization in this country, according to some, would be inconceivable if we used only half as much energy as now. But that is just what we did use in 1963 when we were at least half as civilized as we are now.

See how many forms and sources of energy you can identify that flow into and out of your home in various forms in the course of a day. Do you see any ways to reduce your net use of energy (inflow minus outflow)? Could you live comfortably with only half your present use, as Lovins mentions?

Think about this . . . *Would it save energy to extend daylight saving time throughout the year? If so, should we do so? Why do we use it now?*

Issue 59

Solar Energy

The energy of the sunlight falling on the roof and sides of your car on a sunny day is almost as great as the energy delivered by the engine! And it's free!

The yearly average of sunlight on a horizontal surface when the sun is shining is:

Connecticut 0.2 hp/m^2
New Mexico 0.3 hp/m^2

At 96 km/hr (60 mi/hr), an ordinary car engine doing 13 km/liter (30 mi/gal) consumes energy at the rate of 100,000 W (watts) (134 hp [horsepower]), turning 80,000 W (107 hp) into wasted heat and 20,000 W (27 hp) into mechanical work.

What are the prospects of a gasless car running successfully in New Mexico? In Connecticut? If it could run, why isn't such a car in general use?

(A car that drove across Australia on solar energy is described in "Fill 'er Up with Sunlight," W. H. Jordan, *Smithsonian,* Feb. 1988 and "The Lessons of Sunraycer," H. G. Wilson, P. B. MacCready, and C. R. Kyle, *Scientific American,* March 1989.)

Issue 60

Energy Conversion: Developing Countries

In the farming areas of developing countries, human muscle power is often the most practical source of energy for producing electricity. For operating radios and televisions, battery packs and solar cells are prohibitively expensive, and small gasoline engines driving generators are unreliable and hard to maintain. But a person pedaling a bicycle can easily generate 35 watts, or about 0.05 horsepower (746 W = 1 hp), and a dynamo driven by someone pedaling a stationary bicycle is fairly easy for untrained people to repair.

If your input from food is 2400 kilocalories per day or 100 kilocalories per hour, your input from food, averaged over the day, is 420,000 joules per hour, which is 117 joules per second or 117 watts. Hence, your efficiency at pedaling a simple electric generator is

$$\frac{\text{output}}{\text{input}} = \frac{35 \text{ W}}{117 \text{ W}} = 0.30 \text{ or } 30\% \, ,$$

which is better than a gasoline engine for the length of time that you or a farmer can keep pedaling—presumably for hours.

The same pedal-power applied to a stationary bicycle can be rigged to pump water out of a well or to drag a plow across a field with a winch or to operate winnowing or grinding machines.

Sending powerful labor-saving machinery to developed countries may seem like a good idea, but it usually is not. The 18,000 tractors once shipped to Pakistan increased no crop yields there but simply put thousands of people out of work. Considerations like these guide the development of what is called "appropriate technology" for helping people in developing countries that have little or no access to Western technology but have a large under-used labor force.

People from developed countries working overseas use elementary physics in a similar way:

— to improve pumps for getting clean water from deep underground,
— for building small cement plants and brick-making plants,
— for designing simple metal-working machinery, and
— for inventing more efficient windmills and more effective farming tools.

All use locally available materials. Each developing country has different needs and different solutions—different appropriate technology.

Suppose it takes 10,000 laborers two years to build an earthen dam in India using shovels and wheelbarrows. Suppose also six American bulldozers could do the same job in three months. If you had to make the decision (and some person has to), would you send the bulldozers to India as part of our ongoing foreign aid? If not, why not? Is your answer an argument against giving American technical aid to developing countries? What about sending bicycles instead of gasoline-driven vehicles?

Perhaps you will become interested in using your knowledge of science in this practical way abroad.

"Oh, good—heavy boxes. They'll hold down the thatch on our roof in those torrential winds."

Earth and Space Science

The knowledge of the world is only to be acquired in the world, and not in a closet.

— EARL OF CHESTERFIELD

Issue 61

Earthquakes

Scientists slowly are increasing their ability to predict earthquakes. If you are a scientist who has just determined that there is a 90 percent chance of a major earthquake destroying a town in the next 24 hours, what should you do? What should the public-safety authorities do? What if it was only a 50 percent chance? What should you and the authorities do if there is a 90 percent chance of the earthquake occurring sometime during the next month?

If you evacuate the town and there is no earthquake, what consequences do you anticipate? If you fail to evacuate and the earthquake occurs, what then? Where is the "break even" point?

In 1980, when Mt. St. Helens was becoming active, the immediate area was evacuated. Normal sawmill and lumbering activities involving approximately 10,000 people continued at what was considered to be a safe distance from the volcano. On Sunday, May 18, at 8:32 A.M., an explosive eruption blew off 12 cubic kilometers (3 cubic miles) of earth and created a 5.1 earthquake on the Richter scale, obliterating a large fraction of the area of lumbering and sawmills. Many of the work force would have been killed—except that it was a Sunday so they were not at work. A judgment call that "lucked in"!

San Francisco has been built on top of a fault along which two plates in the earth's crust meet. As the plates move horizontally past each other, rock in some places along the fault gets stuck and stops moving. The force between the plates builds, and an earthquake becomes more and more likely—and more severe when it finally comes. It has been suggested that if a small nuclear explosion could be detonated deep within the earth, the force between the plates could be relieved, thus saving the city and its people from a future great catastrophe.

Would you advocate a detonation? Why? Who should make the decision to detonate: the city, state, or federal government? Popular vote?

Issue 62

American Space Program

America's space program is bound to have important effects on the taxpayer (you), on science and industry, on our foreign policy, and on our collective imaginations, energies, and national character.

1. What should be the goals of our space program?
   a. Construction of a habitable Earth-orbiting space station
   b. Construction of a permanent moon base (to do what?)
   c. Human exploration of Mars
   d. Earth-orbiting satellites to monitor the Earth-surface environment
   e. Search for rare or valuable minerals on the moon or asteroids
   f. Other
2. Should the goal be international or exclusively American? Why?
3. If "international," should the U.S. and the U.S.S.R. cooperate?
4. Is our present radio astronomy "Search for Extraterrestrial Intelligence" (SETI) a worthwhile expense of taxpayers' money?
5. Is it important the U.S. remain in the forefront as a space-faring nation?
6. Should taxpayer money now being spent on space projects be sharply reduced and put to "better" use here on the earth?
7. Should we pursue military uses of space?
   a. "Star Wars" (SDI) is at best a practical shield against nuclear surprise attack and at worst an impractical "make work" project to create jobs and lucrative contracts at taxpayers' expense. Should it be continued? Slowed? Stopped?
   b. Do surveillance satellites provide stability to volatile international political situations? In 1961, satellite photographs of Soviet missiles and bomber facilities dispelled the idea of a U.S.–Soviet missile gap and averted an expensive U.S. catch-up program.
   c. Would U.S. nuclear missiles orbiting the earth and available for instant call-down enhance our security or diminish it?

Issue 63

Outer Space: Infinite Dump

The universe is so vast that we could never fill it with garbage. Therefore, it might serve as a dump into which we could launch such inconvenient trash as spent nuclear power plant fuel rods, radioactive waste, and troublesome chemicals, all of which now pose serious disposal problems and pollute the earth. Or we might shoot it all into the sun, the ultimate incinerator. Why not do either?

Expense and the hazards of a launching rocket failure are two reasons why not. A third is that there already is a lot of trash in earth orbit, creating a problem. It is estimated that in 1988 there were at least 40,000 orbiting objects ranging from scattered paint flecks up to intact satellites flying endlessly overhead, 7100 of them bigger than a baseball.

The space shuttle runs only a one-in-a-million chance of hitting something on each of its orbits. If it hit an ordinary iron bolt at a typical collision velocity of about 10 km/sec, the impact would be the same as being hit by an exploding hand grenade.

Much of the orbiting debris comes from military anti-satellite tests. There will be a great deal more orbiting debris if Strategic Defense Initiative (SDI, or "Star Wars") experiments result in the fragmentation of many more orbiting objects. Particularly vulnerable is the $1 billion Hubble space telescope, whose fragile mirrors and computers could be destroyed by objects a few millimeters across. There are no worldwide rules about what can or cannot be sent into orbit. What problems do you see in cleaning up the space junk (like roadside litter)? Or restricting further launches of vulnerable spacecraft? Or using junk-free orbits farther out in space?

Orbiting debris is already responsible for false astronomical "discoveries." What were thought at first to be pulsing stars sending out powerful but rare optical flashes have turned out to be reflections of sunlight off the solar panels of dead tumbling satellites.

There are other serious threats to astronomy. Worried astronomers already have attempted to block a Melbourne, Florida, company from orbiting the compressed ashes of cremated humans aboard small but shiny space-age mausoleums. It would cost only slightly more than $8 million to orbit a reflective sail one-third the size of a football field that would exceed the brightness of the full moon and blot out many objects important to astronomers.

Seven-figure advertising budgets that now capture a mere national audience could, if channeled into orbiting ads of various kinds in space, make the entire world a captive audience for years. The economic incentive to do so will be enormous. The pressure on countries (and even religions) to colonize "Heaven" with their messages will likewise be great. Presently, no international agreements exist to prevent the orbiting of bright political or religious or commercial symbols, including luminous golden arches advertising McDonald's hamburgers. (See "Scattering of Light: Light Pollution," p. 89.)

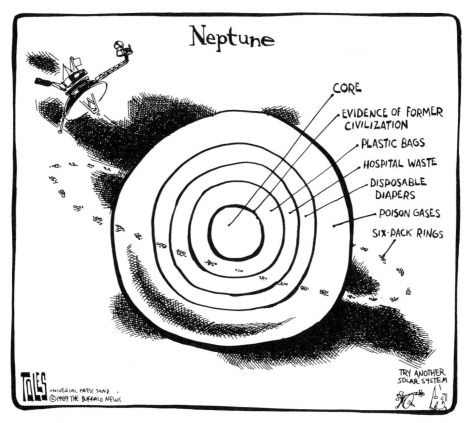

Tom Toles, © 1989 The Buffalo News.

Issue 64

Extraterrestrial Intelligence

If extraterrestrial intelligence (ETI) finally is discovered elsewhere in space and we have established communication with it through our radiotelescopes and receivers, what questions do you think we should ask of it? What information do you think we should send out about ourselves? What about pictures? Music?

Think about this . . . In the 13th century, the sonnet was invented in Sicily and the button in northern Europe. Given the different climates and resultant lifestyles, why is it unlikely the northerners would have devised a new form of poetry or the southerners an efficient way to fasten clothes?

"Men have become the tools of their tools." H. Thoreau, Walden. *True?*

Social Science

My fellow Americans: ask not what your country can do for you—ask what you can do for your country. My fellow citizens of the world: ask not what America will do for you, but what together we can do for the freedom of man.

— John Fitzgerald Kennedy

Issue 65

Tragedy of the Commons

When everyone looks out for his or her own interests and no one looks out for the interests of the entire community (the "commons"), all eventually suffer. For example, one family might discharge their sewage into an unpolluted river, using the reasoning than "the little bit of pollution we contribute can't hurt that much." But if everyone in town did the same thing, the general health and well-being of the public downstream would deteriorate.

Another example: The splendid sand dunes at the Cape Cod National Seashore are being destroyed by the ravages of human feet and tire treads. Efforts by the National Park Service to restrict climbers on the dunes led one member of the Provincetown Board of Selectmen to exclaim, "It seems like every year they come up with more ways to deprive people of recreational activities. You can't take your dog out there; you can't pick the flowers. What are the dunes for? You can't admire them if you can't get on them."

Standards and laws are enacted in many situations to prevent nearly harmless individual acts from escalating to public disasters. For example:

— Factories creating air pollution
— Running in the school corriaors
— Highway speed limits
— Throwing trash along the highways and in the school lunchroom
— Exhaust emission standards for cars
— Toxic waste dumping
— Aircraft traffic control and highway traffic lights
— Hunting whales
— Fishing out of season in local streams and ponds

Similar cases for which few or *no* restrictive laws exist:

— Large-scale fishing on the Grand Banks, the North Sea, and other productive ocean waters
— Having one more baby in the family
— Driving one more car into an air-polluted city
— Paving over one more acre of farmland for a housing development
— Cutting trees in the world's remaining tropical forest

(See "Destruction of Tropical Forests," p. 33.)

Would you support laws restricting any of these activities? Which ones?

All goods and practices involving public or group ownership, such as the atmosphere, the oceans, our parks, and our natural environment—the commons—are examples of resources vulnerable to being overspent. In a commons, the gain to each individual user is exclusively a private gain *in the short term*. The resulting environmental degradation is spread out among *all* the users and only shows up *as a community loss in the long term,* when the resource shows severe stress or signs of collapse.

How many examples of over-used, severely stressed community possessions can you identify in your home, in your community, in the U.S., and in the world? (Many are scattered throughout this book.) As one example: Should dune buggies and other off-road vehicles be allowed in the dune or desert areas of national parks and other public lands where they destroy the fragile vegetation—and the silence?

Garrett Hardin has said, "Freedom in a commons brings ruin to all." Do you agree? (See the classic essay, "The Tragedy of the Commons," Garrett Harden, *Science,* 1968 (162) 1243-48, and reprinted frequently elsewhere.)

(See "Ocean-Bottom Minerals," p. 66 and "The Geopolitics of Genes," p. 18.)

Think about this . . . *When astrology leads people to base serious economic and political decisions on its predictions, it is no longer a game. Isn't it dangerous?*

Issue 66

Exponential Growth

Exponential growth is deceptive because nothing seems to happen for a long time and then the situation explodes dramatically. Here are two examples.

The Sunday newspaper is 2.5 centimeters (1 inch) thick. Fold it in half, and it's 5 centimeters thick. Imagine you could go on folding it indefinitely. How many times would you have to fold it to make it become 1,500,000,000 *kilometers* (93,000,000 miles) thick, the distance to the sun? Answer: Just 43 times. The last fold would add half the total distance. By contrast, 43 steps of *linear* growth would make the paper only 112 *centimeters* (44 inches) thick.

Again, would you rather be paid a million dollars a day for a month or receive 1 penny the first day, 2 the second, 4 the third, 8 the fourth day, and so on, doubling each day—an exponential rate of increase? If you choose exponential growth, you receive only $5.12 on the tenth day, but on the twenty-eighth day you outstrip your million-dollar-a-day colleague. If you went on to day 44, you would receive all the money in stock in the entire U.S., and by day 50 you would own the entire wealth of the world.

In real life, population growth is exponential in many countries. The doubling time for a population growing at the rate of 2 percent a year is 70/2 = 35 years. In general, for any quantity growing exponentially, the doubling time in years = 70/percent growth per year. Thus, if electric power use is growing at 5 percent per year, a seemingly harmless rate of increase, it will have doubled in 70/5 = 14 years. This formula also applies to your money earning 5 percent in a bank savings account.

World population is growing at the rate of 2 percent per year. This means we must double world food production in 70/2 = 35 years simply to hold constant the fraction of the world's population that is starving.

Modern agriculture is based on petroleum-derived fertilizers and petroleum-powered farm machinery. (In the U.S. 750 liters [200 gallons] of gasoline or its equivalent are used to raise 1 hectare [2½ acres] of corn.) But world reserves of petroleum are being depleted exponentially. As exponentially exploding populations press against exponentially declining and irreplaceable natural resources, the world is steadily losing its capacity to feed itself.

How much longer may our petroleum, coal, and iron resources last? How much longer may a community's underground water supply (yours?) keep up with an exponential growth of population? How much longer may commercial whaling continue? Whatever the answer, one thing is certain if the use of the resource is growing exponentially: whatever the limit, half of the entire resource is used up in just *one* doubling time—the last one. This mathematical certainty means that the depletion of a fixed natural resource will lead to a sudden crisis unless foresight and long-range planning are supported by political and economic action long before that painful last doubling.

Can you find examples of such foresight? Or examples of foresight that have led to effective planning?

(See "Petroleum Depletion," p. 65 and "Tragedy of the Commons," p. 110.)

Resources about to be exponentially depleted.

Issue 67

Feedback: Positive and Negative

We live in a world of constant change. Fluctuations of an animal population, economic cycles, and the varying temperature inside your home are a few examples. Some situations stabilize themselves and level out because the fluctuations set in motion some corrective action, called *feedback,* that reduces the fluctuations.

For example, the population of wild rabbits regularly fluctuates between scarcity and abundance. One reason this occurs is that rabbits are important in the diet of foxes. If there is an abundance of rabbits, the foxes get plenty to eat, and they multiply. But with more hungry foxes out hunting, the rabbit supply becomes depleted. Foxes begin to starve, With fewer foxes, the rabbit population can increase once again. Thus, an *increase* in the rabbit population leads to an increase in foxes, which tends to *reduce* the rabbit population—a self-correcting mechanism, called *negative* feedback, controlling the rabbit population.

Can you explain why the following are also examples of negative feedback?

— A thermostat controls the temperature of a room.
— Our economic system has slumps and booms. (Take heart! No one has fully explained this one.)
— A person can stand upright without falling over.
— A passenger in a speeding car says, "Too fast, slow down!" whenever speed is too great.
— The rate of a human heartbeat is controlled.

Some unstable situations fail to stabilize themselves naturally because any fluctuations that occur set in motion an action that *increases* the fluctuation. You can demonstrate such a *positive* feedback situation by putting a microphone attached to an audio amplifier in front of the loudspeaker output. The faintest sound picked up by the microphone is amplified, and the amplified sound coming out of the loudspeaker is picked up and amplified further, ad infinitum. What limits the final loud noise? (Answer: The power rating of the amplifier.)

As another example, when sunlight melts snow in the Northern Hemisphere, the dark background exposed under the snow absorbs more heat, causing additional warming and heating. This positive feedback hastens the onset of spring in the snowbound North.

The spread of a forest fire is also a consequence of positive, not negative, feedback. So is the behavior of a crowd in a theater when someone shouts, "Fire!" Or, when CFCs used as refrigerants escape into the atmosphere, they induce a climatic warming that increases the need for CFC-powered air conditioners, the use of which may further deplete the stratospheric ozone layer and increase further the need for the product.

Can you explain why an arms race may be yet another example of positive feedback?

Try to identify other examples of feedback, positive or negative, in your daily life.

Consider the more complicated feedbacks at work globally between energy consumption, agriculture, and the greenhouse effect.

> ***Think about this . . .*** *Disease, abortion, wars, murder, famine, accidents, and pollution reduce the world's population. Its* increase *is promoted by motherhood, medicine, public health, peace, law and order, scientific agriculture, accident prevention, and clean air. Therefore, why isn't the population explosion desirable?*

Issue 68

Long-Term Effects of Technology

Human activities are affecting the environment in ways we do not yet understand because a generation of human lives is too short a time for us to determine:

— the effects on the atmosphere of **carbon dioxide** from automobiles, power plants, and factories (the greenhouse effect).

— the possible genetic effects of **radiation** from the burning of coal and from the use of nuclear power plants.

— the effects on humans of **drugs and chemicals** of all kinds.

— the long-term effects of **birth control** on populations.

— the long-term consequences of our **destruction of plant and animal species** and our worldwide **destruction of forests**.

Whatever the effects may be, they will be shaped to a great extent by actions we take today. In each case, what actions do you see taking place? What do you think our actions *should* be?

"Industrial Evolution"

Drawing by Andy Lemons.

Issue 69

Science and Congress

About half the bills presented to the Congress require some knowledge of science for their full understanding. Have your congressmen and congresswomen studied much science? If they have not, how do they go about deciding how to vote? Should your congresspersons be required to show that they have had (and passed) some high-school level science courses before they take office?

> ***Think about this . . .*** *Why isn't the proven technology of cloud seeding used to produce rain in drought-stricken areas?*
>
> *Should anthropologists be permitted to exhume the bones of deceased Indians in order to obtain scientific information? Even when their tribal descendants protest?*

Issue 70

Women as Scientists and Inventors

A National Inventors Hall of Fame near Washington in 1984 had 52 members. Not one was a woman. Not:

— **Marie Curie,** who discovered radium and won a Nobel Prize in physics in 1903.

— **Lise Meitner,** who discovered and named nuclear fission.

— **Gladys Hobby,** for the development of oxytetracycline (Terramycin™).

— **Carrie Everson** for her oil flotation system for separating gold and silver from rocky ore.

— **Eleanor Raymond** and **Maria Telkes,** for major contributions to modern domestic solar heating.

— **Jocelyn Bell,** who shared in the discovery of the first pulsar.

— **Maria Goeppert Mayer,** who won a Nobel Prize in physics in 1963 for her work on the shell model of the atomic nucleus.

— **Bette Graham,** who invented Liquid Paper.

— **Rosalind Franklin,** who shared in the discovery of the double-helix structure of DNA.

— **Barbara McClintock,** who won a Nobel Prize for her demonstration of the transfer of genetic information among a single individual's chromosomes.

This deplorable state of affairs might justify a project to identify more women scientists and inventors and to examine their achievements.

Issue 71

Citizens' Responsibility to Society

What responsibility, if any, do *scientists* have for anticipating or ameliorating the effects of their work on society? An important fraction of all the scientists in the U.S. are involved in military-related research. Does your answer apply equally to them?

Is the *engineer* responsible for the consequences when his or her product fails (e.g., automobile manufacturers for faulty design of your car's brakes, or a chemical company for the carcinogenic effects of dioxin in its herbicide)?

Is the *legislator* responsible for failure to legislate safety standards (e.g., not requiring seat belts when most other states do—so you are killed in a collision, or failing to fund the clean-up of a probably hazardous waste dump)?

Is the voting *citizen* ever responsible for the consequences—good or bad—of his or her vote?

Are *you* responsible in any way for the consequences of your job? Suppose you find a well-paying job working on atomic weapons or building tanks for the military?

Many scientists and engineers have refused to take such military-related jobs. Would *you* do so? Do you approve of "whistle blowers"—workers in government or industry who, as an act of conscience, report shoddy work or dishonest practices by their fellow workers?

Think about this . . . *"Scientists should be on tap but not on top." Winston Churchill. Do you agree?*

Issue 72

Risk

Is it possible to create a risk-free society? Identify risks you could eliminate from your own life by making personal choices.

Every day as a matter of habit, people continue to kill themselves by smoking cigarettes and driving without seat belts while at the same time being afraid of the relatively smaller health risks of artificial sweeteners, salt, and cholesterol. Unthinking habit, perhaps, but in other cases the decision to act or not to act required a conscious weighing of the consequences.

For example, in the spring of 1986, the European tourist trade collapsed due to the hijacking of the cruise ship, *Achille Lauro,* the bombing of a TWA jetliner over Greece, and the confrontation with Libya. People consequently feared traveling abroad. But during the entire preceding year only 23 Americans had been killed anywhere overseas by terrorists—not an unusual number— while close to 45,000 Americans were killed in motor-vehicle-related accidents here in the U.S. Would you have canceled or postponed a trip to Europe or the Middle East in the spring of 1986? Why?

As another example, two independent teams of scientists agree that the southern third of the San Andreas fault extending past San Bernardino, California, has a 25 percent chance in the next 20 to 25 years of having an earthquake as severe as the one that devastated San Francisco in 1906. What effect might this announcement have on real estate values and insurance rates in the area? Would you accept a good job that required you to live in the area?

If you are a responsible public official who has just learned from reputable seismologists that there is a 50 percent chance of a major earthquake destroying your town in the next few days, what should you do? What are the consequences if you do not act and the earthquake occurs? What happens if you do act and the earthquake does not occur?

Radon, an invisible, odorless, radioactive gas is produced by the radioactive decay of uranium in soil and rocks in scattered areas of the U.S.

When radon accumulates in poorly ventilated homes, its radiation can result in lung cancer. It is generally accepted that radon is responsible for about 10,000 of the nation's 130,000 annual lung cancer deaths. If you find that your home's basement has an unusually high concentration of radon gas, you or some other responsible person has to decide how much money to spend on renovation of your home. No one knows the probable cancer risk at a given level of radon exposure, but your failure to act will unquestionably increase your odds of getting lung cancer. Though you can reduce the risk, you can never eliminate it.

At what level of risk should the government subsidize corrective action in the homes of the impoverished? How do you feel about the construction of a nuclear power plant upwind from your town? Or about the radioactive waste and hazardous industrial chemicals that pose a danger to the communities through which it must be transported? Can you defend your opinion with reasons? (See "Radioactive Waste Disposal," p. 59.)

We frequently put values on our lives, as when we buy a lightweight car instead of a more expensive gas guzzler or decide not to install an expensive safety device as protection against an unlikely category of accident. Which would you rather have: an entirely safe job at $200 per week or a job at $1000 per week with a 1-in-10 chance of coming down with cancer within 20 years?

Consider this example: The spacecraft *Apollo 13* was launched from Cape Kennedy Space Center pad 39 (the third multiple of 13) at 1313 hours central time on its way to the moon. On the 13th of April, 1970, an explosion forced the craft to return to the earth, having barely escaped destruction. As the NASA official in charge of scheduling spacecraft and shuttle launches in the future, would you authorize a launch at 1313 hours on a Friday the 13th if the extreme pressures on the timing of launch made this the most appropriate moment? Why or why not? Remember the *Challenger* disaster!

Is it unethical for doctors or dentists to refuse to care for people infected with the AIDS virus for fear of contracting the disease themselves? Many doctors who do so argue that since the disease is *invariably* fatal, *any* risk is too high. Other doctors argue that many serious diseases are much more communicable.

Identify half a dozen hazards you consider most threatening to your health or safety. Try to rank them by probability. Is it practical to reduce their probabilities to zero?

Issue *73*

Metric System

Why bother changing over to the metric system?

Advantages:
1. It's much simpler to use. You change kilometers to centimeters by merely moving a decimal point. If you doubt the ease, try changing miles to inches!
2. It's used in every other nation in the world except Burma, South Yemen, and Brunei. Hence, it's important that our industries use it in manufacturing for export and competition in the world marketplace. General Motors is almost entirely metric.
3. Pharmaceuticals and photographic materials are specified in metric units. Nutrition information shows a strange mixture of metric and U.S. customary units. Many sports events use metric specifications. News of the drug war constantly familiarizes the public with metric mass measures.

Disadvantages:
1. The public resists strongly; some people have even organized to fight what they see as a troublesome, even impractical, move. For example, deeds, plats, and blueprints on record for hundreds of years back are all in inches and feet. Consumer advocates warn that a changeover to metric packaging in the marketplace will provide a smokescreen to hide steep price increases.
2. In September 1982, after six years of effort, the federal government gave up exhorting the country and abolished the U.S. Metric Board. The new, much smaller Office of Metric Programs assumes that eventually, perhaps not in your lifetime, enough people will be used to metric measurements so that the change will take place slowly and all by itself.

3. What metric units do you feel comfortable using already: kilograms, meters, liters? Should stores selling such consumer goods as food, hardware, and clothing be encouraged to use them?

4. Isn't it odd that people are so enthusiastic about learning to use computers, which is much more difficult than learning to use the metric system?

5. While most measurements are being converted into metric units these days, the football field will probably always remain 100 yards long. Not that no one has tried to change it. *Chemistry* magazine ("Drop Back Ten Meters and Punt," Nov. 1977, p. 4) reports an experiment at Carleton College in Minnesota with a field 100 meters long and 50 meters wide (28 feet longer and 14 feet wider than usual). In the experimental game with its rival, St. Olaf, Carleton lost 43-0. It seems that the experiment was abandoned.

6. A metric-based index relates your body mass to your health. You can calculate your own index by dividing your mass in kilograms by the square of your height in meters. For values of the index above 25, there is a steady decrease in your life expectancy and an increase in the likelihood of certain diseases. (*U.S. Metric Association Newsletter,* July-August 1985, p. 7).

(See "Alphabetization: Chinese," p. 128.)

METRIC CLOCK

Issue 74

Computers and the Workplace

Will the widespread use of computers of various kinds eliminate the need for central offices?

Workers whose jobs are done entirely on computers might just as well have their terminals at home, tied into the company's central computer by a telephone link.

Think of the effects on our big cities if workers could do just that: fewer commuters; more relaxed rush-hour traffic; a decline in office real estate, including the need for big central-city skyscrapers; an accelerated destruction of farmland as suburban commuters move farther out into the country; households in which the breadwinner is home all day long; the isolation from gossip and office politics; and the loss of a sense of community with fellow workers.

Will we find that the human need for society is so strong that these dispersals will be resisted?

Given the advantages and disadvantages, would you choose to spend your working life at such a "home work station"?

The athletes at many Olympic events could perhaps perform better without leaving home. There, timed and measured electronically, their performance could be communicated immediately to central headquarters, complete with pictures, from all over the world, eliminating immense expense and controversy. Would this be desirable?

(See "Automation Costs Jobs," p. 125.)

Think about this . . . If a woman has a legal obligation to provide prenatal care for her child, how can she at the same time have the right to abort the fetus?

Issue 75

Automation Costs Jobs

Research at some American colleges and universities is leading to the development of labor-saving machinery for use on farms. In recent years, huge machines that harvest crops have eliminated tens of thousands of jobs and have contributed to the growth of ever-larger corporate farms and the elimination of small farmers.

Research in other laboratories has led to the development of the transistor and then the microchip, which in computers has automated factories and offices and eliminated innumerable jobs.

Robots are being used more and more in manufacturing industries (e.g., automobiles and refrigerators), both in the U.S. and abroad. Unfortunately, they too replace people and, therefore, cost jobs. But highly automated plants raise productivity, thus offsetting cheaper labor costs in other countries. Their untiring precision makes possible a uniformly high quality of product that is economically competitive in world markets.

Such developments as these raise the specter of what has been called "jobless economic growth."

Would you be willing to take a job developing a machine or a device that you knew would, if successful, put people out of work? For example, how about a robot to drive farm machinery?

What if the displaced worker were a member of your own family?

If you had been a member of the California legislature when a bill came before it to suspend state support for research and development of agricultural labor-saving machinery because their use would destroy farm workers' jobs, how would you have voted? (The measure failed.)

(See "Computers and the Workplace," p, 124.)

Issue 76

Computers and Privacy

"**B**ig Brother" *can* watch you!

Two-way cable television technology allows a viewer to respond to questions or to initiate a variety of transactions by pushing buttons on a small hand-held console connected from a TV set to the cable TV company headquarters. Early versions tried out in 1984 in some cities (e.g., Pittsburgh, Cincinnati, Knoxville, and Syracuse) could be expanded to provide electronic banking; shopping; opinion polling; meter reading; burglar, fire, and medical emergency protection; and education and entertainment.

Because it is now possible to store enormous amounts of information in a tiny silicon computer chip, it is easy to save information fed into a television company's headquarters by its cable subscribers. As a result, a subscriber's two-way television set becomes, in effect, a guest that puts on file at headquarters the person's comings and goings, banking transactions, opinions on controversial issues, taste in consumer goods, and the viewing preferences of the household. These data are a gold mine of confidential information that the cable TV company may sell to advertisers, politicians, sociologists, or gossip columnists.

The right to privacy is a cornerstone of liberty. To the degree that subscribers choose to take advantage of this technology, it may be argued that it places the liberty of its users in the hands of every two-way cable TV operator.

While Americans may bristle at the idea of identity numbers, they accept the reality. Every time they change jobs, they readily submit their Social Security numbers to ensure that employer contributions are credited to them. Every time they use an electronic ticket-ordering service or call an 800 number for mail-order moccasins, they give a credit card number. Should the use of such identification systems be regulated? If so, how?

Of course, you broadcast your personal conversations to the world every time you use a cordless phone.

Does this technology pose a serious danger to you? Should any action be taken? If so, what action? By whom?

Another example from a report on March 1, 1989, in *The New York Times:* The Nynex Corporation will offer customers a "caller identification system that would display the number from which an incoming call was originating before the call was answered." This system would allow customers the advantage of screening their calls or identifying the sources of obscene or crank calls. It would be disadvantageous to callers wishing to protect their unlisted telephone numbers or their identity from a crisis hotline. Should the use of such caller identification systems be regulated? If so, how?

Issue 77

Alphabetization: Chinese

In 1956 the Chinese government began to convert Chinese writing to our familiar 26 letters. It simply was too inefficient to alphabetize everything from telephone books and dictionaries to tax registers by the 214 radicals used to write their complicated-looking characters or ideograms.

If we had 214 letters in our alphabet, how much longer would it take you to learn to write or to look up a word in a dictionary or a name in a telephone book? Could we do with fewer letters in our own alphabet? Would the advantage of fewer letters outweigh the problems created by trying to eliminate some of them? Compare your answers with the present efforts to replace our British units of measurement by the metric system.

Word processors have been devised that can convert entire sentences written in Chinese characters into grammatically correct Japanese or Korean or vice versa. This invention will change the fabric of Asia's cultural and economic life in fundamental ways. It will bring more women into the workforce (who master the machines more quickly than do their male counterparts), speed communication in the business world, bring together diverse cultures previously divided by language barriers, and raise the level of literacy of Asian nations.

An even more dramatic invention is the telefax or facsimile ("fax") machine, which transmits pictures and handwritten messages, as well as typed and printed material. What special effects might the fax machine be expected to have on commercial transactions, political life, and the standard of living in Asian countries? (See "Metric System," p. 122.)

Think about this . . . *Taxpayers' money is presently used to search for life elsewhere in the universe. Should this effort be stopped? Expanded? Left unchanged?*

Issue 78

Popular Ideas about Science

To compare your attitude toward science with the views of 2000 American adults tested in 1985 by Jon D. Miller, head of the Public Opinion Laboratory at Northern Illinois University, answer the following questions (true or false).

1. Some numbers are especially lucky for some people. (40)
2. Because of their knowledge, scientific researchers have a power that makes them dangerous. (over 50)
3. The positions of the sun, moon, planets, and stars have at least some influence on human affairs (astrology). (40)
4. It is likely that some of the unidentified flying objects (UFOs) that have been reported are really space vehicles from other civilizations. (43)
5. Space shots have caused changes in our weather. (over 40)
6. Science tends to break down people's ideas of right and wrong. (over 30)

Of the 2000 test takers, the percentage answering "true" is in parentheses after each question. Ideally, the numbers should all be zero (0) since *all the statements are false*. Assuming the sample was a fair cross section of adult Americans, what do these numbers show about attitudes toward science? Compared with them, how well did you do?

In another poll reported in *The American Biology Teacher* (editorial, May 1989) 34 percent of those polled thought psychic powers could be used to read people's thoughts, 29 percent felt we could communicate with the dead, and 22 percent believed in ghosts. The sample being polled consisted of high school life science and biology teachers!

Issue 79

Astrology

A Gallup poll in 1984 indicated that 52 percent of American teenagers believed that astrology works. Astrology columns appear in over 1200 newspapers in the U.S. Does this mean that astrology should be taken seriously? Consider some of the evidence and some pointed questions:

1. What is the likelihood that one-twelfth of the world's population is having the same kind of day?
2. Why is the moment of birth, not conception, crucial for casting a horoscope?

"The practice of astrology took a major step toward achieving credibility today when, as predicted, everyone born under the sign of Scorpio was run over by an egg lorry [truck]."

Drawing by Bud Grace. Reproduced by permission of Punch.

3. If astrologers are as good as they claim to be, why aren't they richer?
4. Are all horoscopes that were done before the discovery of the three outer planets incorrect?
5. Why do different schools of astrology so strongly disagree with one another?
6. If astrologers claim that your personality is dictated by the attraction of the planets at the moment of your birth, how do you account for the fact that the gravitational pull of the delivering obstetrician far outweighed that of the planets involved?

After the San Francisco earthquake in October 1989, *Washington Post* columnist Richard Cohen found that Nancy Reagan's much-publicized astrologer had been home in San Francisco. "What I want to know is why a woman who told the president the precise moment to sign a treaty couldn't see an earthquake coming," he wrote. Alas, she said, that kind of prediction is best left to an earthquake astrologer. ("In Brief," *Skeptical Inquirer,* Spring 1990, p. 239.)

Should people who believe in astrology be allowed to teach science in public schools, practice medicine, or hold public office? Would you vote for such a person if you had the choice?

Think about this . . . *Would you be willing to use a drug or cosmetic that you knew had been tested extensively on live animals? Would you use a drug or cosmetic that was guaranteed not to have been tested on live animals and might therefore be unsafe?*

High school football, power mowers, and mountain climbing take more lives each year than nuclear power plants. Should they be abolished before nuclear power plants are?

Issue 80

Science and Non-Science

Which of these items are supported by scientific evidence? Which are supported by popular tradition with little or no scientific evidence? How do you tell the difference?

— Bermuda triangle
— extra-sensory perception (ESP)
— acupuncture
— homing instinct in pigeons
— astrology
— prediction of eclipses
— astronomy
— flat (or hollow) earth
— biorhythms
— continental drift
— UFOs
— pyramid power
— creationism
— evolution of species

Think about this . . . *Did it violate animal rights to use a baboon's heart for a human heart transplant (Baby Fae, 1984)? If so, what about the steer that provides your hamburger for lunch?*

Appendix

The optimist proclaims that we live in the best of all possible worlds; and the pessimist fears this is true.

— JAMES BRANCH CABELL
The Silver Stallion

Appendix

Quotations 1

Science as Foolishness

1. In debate during February 21, 1861, on an item that would appropriate $6000 to the Smithsonian Institution, Senator Simon Cameron said, "I am tired of all this thing called science here. . . . We have spent millions on that sort of thing for the last few years, and it is time it should be stopped."

2. "We hope that Professor Langley will not put his substantial greatness as a scientist in further peril by continuing to waste his time, and the money involved, in further airship experiments. Life is short, and he is capable of services to humanity incomparably greater than can be expected to result from trying to fly . . . For students and investigators of the Langley type there are more useful employments" (*The New York Times,* Dec. 10, 1903, editorial page).

3. "The demonstration that no possible combination of known substances, known forms of machinery, and known forms of force can be united in a practical machine by which man shall fly long distances through the air, seems to the writer as complete as it is possible for the demonstration of any physical fact to be" (Simon Newcomb, eminent American astronomer [1835-1909] quoted by Arthur C. Clarke, *Profiles of the Future* [New York: Harper & Row, 1962] 2–3).

4. "I have always consistently opposed high-tension and alternating systems of electric lighting . . . not only on account of danger, but because of their general unreliability and unsuitability for any general system of distribution.

"There is no plea which will justify the use of high tension and alternating currents, either in a scientific or a commercial sense. They are employed solely to reduce investment in copper wire and real estate" (Thomas A. Edison, "The Dangers of Electric Lighting," *North American Review,* Nov. 1889, pp. 630–633).

5. Aristotelian professors who were contemporaries of Galileo said, concerning his discovery: "Jupiter's moons are invisible to the naked eye, and therefore can have no influence on the earth, and therefore would be useless, and therefore do not exist" (A. Williams-Ellis, *Men Who Found Out* [New York: Coward-McCann, 1930] 43).

6. Criticizing Robert Goddard's pioneering rocket research, a New York Times editorial in 1921 said, "That Professor Goddard with his 'chair' in Clark College and the countenancing of the Smithsonian Institution does not know the relation of action to reaction, and of the need to have something better than a vacuum against which to

134

react—to say that would be absurd. Of course he only seems to lack the knowledge ladled out daily in high schools" (M. Lehman, *This High Man, The Life of Robert H. Goddard* [New York: Farrar, Straus, & Co., 1963] 111).

7. In 1939, U.S. Rear Admiral Clark Woodward said, ". . . As far as sinking a ship with a bomb is concerned, you just can't do it" (Ralph L. Woods, *American Legion Magazine,* October 1966, p. 29).

8. In December 1945, Dr. Vannevar Bush said of intercontinental missiles, "There has been a great deal said about a 3000-mile high-angle rocket. In my opinion such a thing is impossible for many years. The people who have been writing these things that annoy me have been talking about a 3000-mile high-angle rocket shot from one continent to another, carrying an atomic bomb and so directed as to be a precise weapon which could land exactly on a certain target such as a city.

"I say, technically, I don't think anyone in the world knows how to do such a thing, and I feel confident that it will not be done for a very long period of time to come. . . . I think we can leave that out of our thinking. I wish the American public would leave that out of their thinking" (Arthur C. Clarke, *Profiles of the Future* [New York: Harper & Row, 1962] 9).

9. In 1945, Admiral Leahy told President Truman the atomic bomb "is the biggest fool thing we have ever done. . . . The bomb will never go off, and I speak as an expert in explosives" (Harry S Truman, *Memoirs,* Vol. I [New York: Doubleday and Co. 1955] 11).

10. Commenting on the proposal to drive a steamboat by a screw propeller, Sir William Symonds, Surveyor of the British Navy, commented in 1837, ". . . even if the propeller had the power of propelling a vessel, it would be found altogether useless in practice because, the power being applied in the stern, it would be absolutely impossible to make the vessel steer" (W. C. Church, *The Life of John Ericsson* [New York: Charles Scribner's Sons, 1890] 90).

11. In 1913, Lee de Forest, inventor of the audion tube, which made radio broadcasting possible, was brought to trial on charges of using the U.S. mails fraudulently to sell public stock in the Radio Telephone Company, purported to be a worthless enterprise. In court, the district attorney charged: "De Forest has said in many newspapers and over his signature that it would be possible to transmit the human voice across the Atlantic before many years. Based on these absurd and deliberately misleading statements the misguided public . . . has been persuaded to purchase stock in his company. . . ."

DeForest was acquitted, but the judge advised him "to get a common garden variety of job and stick to it" (L. Archer, *History of Radio* [Washington, DC: American Historical Society, 1938] 110).

12. Aristotle maintained that women have fewer teeth than men; although he was married twice, it never occurred to him to verify this statement by examining his wives' mouths. (Bertrand Russell on the importance of experiment.)

13. When I was a boy of 14, my father was so ignorant I could hardly stand to have the old man around. But when I got to be 21, I was astonished at how much he had learnt in 7 years. (Mark Twain)

14. The world was created on October 22, 4004 B. C. at 6 o'clock in the evening. (Irish Archbishop James Usher in 1742.)

Quotations 2

Science in Our Culture

15. Science and technology will play the key role. They will be of decisive significance in the competition between the two systems. (Leonid Brezhnev, Soviet ex-president and general secretary)

16. The splitting of the atom has changed everything save our mode of thinking, and thus we drift toward unparalleled catastrophe. (Albert Einstein)

17. Scientists should be on tap but not on top. (Winston Churchill)

18. That's one small step for a man, one giant leap for mankind. (Neil Armstrong stepping onto the moon, July 1969)

19. Socialism is inconceivable without engineering based on the latest discoveries of modern science. (V. I. Lenin, one of the major architects of Soviet communism; died 1924.)

20. It is science alone that can solve the problems of hunger and poverty, insanitation and illiteracy, of superstition and deadening custom and tradition, of vast resources running to waste, of a rich country inhabited by starving people. (J. Nehru, prime minister of India from the beginning of its independence in 1947; died 1964)

21. The whole of science is nothing more than a refinement of everyday thinking. (Albert Einstein)

22. I have one further observation to make, and that is that you scientists have gotten a long way ahead of human conduct, and until human conduct catches up with you, we are in a precarious way unless you scientists slow up a little and let us catch up. (Senator Johnson of Colorado at hearings before the Special Committee on Atomic Energy. U.S. Senate, 79th Congress, first session)